D1542777

Invariant Imbedding and its Applications to Ordinary Differential Equations

An Introduction

APPLIED MATHEMATICS AND COMPUTATION

A Series of Graduate Textbooks, Monographs, Reference Works

Series Editor: ROBERT KALABA, University of Southern California

No. 1   MELVIN R. SCOTT
  Invariant Imbedding and its Applications to Ordinary Differential Equations: An Introduction, 1973

No. 2   JOHN CASTI and ROBERT KALABA
  Imbedding Methods in Applied Mathematics, 1973

No. 3   DONALD GREENSPAN
  Discrete Models, 1973

Invariant Imbedding
and its Applications to
Ordinary Differential Equations
An Introduction

MELVIN R. SCOTT

Sandia Laboratories

Albuquerque, New Mexico

142138

 1973

ADDISON-WESLEY PUBLISHING COMPANY

Advanced Book Program

Reading, Massachusetts

London · Amsterdam · Don Mills, Ontario · Sydney · Tokyo

CODEN: APMCC

Library of Congress Cataloging in Publication Data

Scott, Melvin R., 1942–
  Invariant imbedding and its applications to ordinary differential equations.

  (Applied mathematics and computation)
  Includes bibliographies.
    1. Differential equations — Numerical solutions.
  2. Boundary value problems — Numerical solutions.
  3. Invariant imbedding.   I. Title.
  QA372.S392   515'.623   73-4792
  ISBN 0–201–06844–3
  ISBN 0–201–06845–1 (pbk)

Reproduced by Addison-Wesley Publishing Company, Inc., Advanced Book Program, Reading, Massachusetts

American Mathematical Society (MOS) Subject Classification Scheme (1970): 65L05, 65L10, 65L15

CONTENTS

SERIES EDITOR'S FOREWORD

Execution times of modern digital computers are measured in nanoseconds. They can solve hundreds of simultaneous ordinary differential equations with speed and accuracy. But what does this immense capability imply with regard to solving the scientific, engineering, economic, and social problems confronting mankind? Clearly, much effort has to be expended in finding answers to that question.

In some fields, it is not yet possible to write mathematical equations which accurately describe processes of interest. Here, the computer may be used simply to simulate a process and, perhaps, to observe the efficacy of different control processes. In others, a mathematical description may be available, but the equations are frequently difficult to solve numerically. In such cases, the difficulties may be faced squarely and possibly overcome; alternatively, formulations may be sought which are more compatible with the inherent capabilities of computers. Mathematics itself nourishes and is nourished by such developments.

Each order of magnitude increase in speed and memory size of computers requires a reexamination of computational techniques and an assessment of the new problems which may be brought within the realm of solution. Volumes in this series will provide indications of current thinking regarding problem formulations, mathematical analysis, and computational treatment.

ROBERT KALABA

Los Angeles, California
April, 1973

PREFACE

What is *invariant imbedding*? The answer to this question is both simple and complicated. Although the actual application of the imbedding is relatively straightforward, the exact form of the imbedding to be used normally must be determined for each new problem. Basically, the method involves generating a "family" of problems by means of a single parameter, where the basic properties of the system remain invariant under the generation of the family. The family then provides a means of advancing from one member, sometimes degenerate, to the solution of the original problem.

In the original applications of invariant imbedding, the generating parameter was the "size" of the system; for example, the length of the interval or the thickness of a slab. In the past few years we have seen that the principles of invariance can take on many different forms. This leads to problems where the imbedding parameter can be any one of several "physical" parameters of the system. Many problems of classical analysis can also be viewed as an "imbedding", where the imbedding parameter is almost always either position in a fixed interval or time.

This book grew out of the author's research in the field over the last half-dozen years. It should be of special interest to engineers, physicists, and applied mathematicians. However, it is not intended to be strictly a research monograph. An early version of the book was used by the author in a one-semester special topics course at the University of Vermont. Also, the level of the book has been designed to be useful as a supplementary text in a course in numerical analysis or ordinary differential equations. A large number of exercise and numerical examples have been included as an aid for tutorial purposes.

In Chapter I some of the basic definitions and theorems of ordinary differential equations are reviewed. This includes material such as Sturm's oscillation and comparison theory for second-order equations. A command of this material is necessary for a thorough understanding of the theory of eigenvalues which is discussed in Chapters V and VIII. Also, a short discussion of Riccati equations is given since they play an important role in the theory of invariant imbedding.

Chapter II is devoted to a review of the classical methods for the numerical solution of initial-value and boundary-value problems. Several forms of instability are briefly discussed.

The first discussion of invariant imbedding is introduced in Chapter III. The invariant imbedding formulation is contrasted to the classical formulation of two problems.

Chapter IV is devoted to the solution of inhomogeneous linear boundary-value problems. The basic variants of invariant imbedding are compared. Several numerical examples demonstrate the efficacy of the imbedding methods.

Homogeneous problems are considered in Chapter V. Methods for computing eigenvalues, eigenfunctions, and eigenlengths for second-order problems with general boundary conditions are presented. Several numerical examples which compare the imbedding technique with other methods are discussed.

Chapter VI reconsiders inhomogeneous boundary-value problems and correlates the results of Chapters IV and V to the classical alternative principle. An algorithm for solving boundary-value problems whose interval lengths are greater than the first characteristic length and two numerical examples are presented.

One of the basic purposes of the invariant imbedding is to convert unstable boundary-value problems into stable initial-value problems. In Chapter VII we show that certain initial-value problems can be converted into stable initial-value problems via an imbedding technique.

Systems of equations are considered in Chapter VIII. One of the real beauties of the method is its natural generalization to higher-order systems. Both homogeneous and inhomogeneous problems are considered. Again, we present several numerical examples to illustrate the efficacy of the method.

In Chapter IX we treat nonlinear boundary-value problems. The classical imbedding equations which are quasilinear partial differential equations with initial conditions are derived. The method of quasilinearization is used to linearize the equations and the resulting sequence of linear boundary-value problems are then solved using invariant imbedding. Several numerical examples are presented.

The author is indebted to a number of individuals. Professors Paul Brock and James Burgmeier of the University of Vermont were instrumental in the basic organization of the book and made a number of excellent suggestions on later versions of the manuscript. Professor Paul Nelson, Jr. of Texas Tech. University, whose work has been included in several sections, read the entire manuscript and suggested a number of changes which materially improved the book.

A special acknowledgment must go to Professor Robert Kalaba of the University of Southern California who has been a inspirational friend and colleague for a number of years. Many of the ideas of this book have grown out of conversations with these individuals over the years.

The author wishes to express his sincere gratitude to Sandia Laboratories and to the U.S. Atomic Energy Commission for their support of research in the development of new numerical techniques. Particular acknowledgment goes to R. L. Coats, J. V. Walker, A. W. Snyder, and L. C. Hebel. Lucile Marcrum of Sandia's editorial staff made a number of corrections and suggestions to the manuscript. Jan Willis directed much of the typing of the manuscript.

This book is respectfully dedicated to my wife, Karen, and children, Vesta, Todd, and Désirée, whose patience and understanding made the completion of this book possible.

MELVIN R. SCOTT

Albuquerque, New Mexico
April, 1973

LIST OF FIGURES

LIST OF TABLES

Invariant Imbedding and its Applications to Ordinary Differential Equations

An Introduction

I
REVIEW OF ORDINARY DIFFERENTIAL EQUATIONS

1 Introduction

We shall review some of the fundamental properties of ordinary differential equations. The primary purpose is to discuss only those topics which will be needed in our development of the method of invariant imbedding. In section 2 the basic existence and uniqueness theorems for initial-value problems are discussed. In section 3 the concept of linear independence and Wronskians is developed. Section 4 is devoted to the classical Sturmian theory which plays a fundamental role in the study of eigenvalue problems. Some of the basic theorems on boundary-value problems are discussed in section 5. Green's functions, which are important to the study of boundary-value problems, are developed in section 6. Section 7 includes both scalar and matrix cases of Riccati equations.

2 Existence and Uniqueness for Initial-Value Problems

A differential equation is an equation which involves certain derivatives of an unknown function and possibly the unknown function itself. Probably the simplest type of differential equation is that given by*

$$\frac{dy}{dx} = f(x); \qquad (I.2.1)$$

that is, we wish to find the function $y(x)$ whose derivation is $f(x)$. In this case

---

* The notations $\dot{y} = y' = dy/dx$ will be used indiscriminately to denote differentation.

1

we need only integrate both sides of (I.2.1) with respect to x to get the solution

$$y(x) = \int_a^x f(t) \, dt + C. \tag{I.2.2}$$

We shall be primarily interested in linear differential equations; that is, equations which can be written in the form

$$\frac{dy}{dx} = A(x)y + f(x), \tag{I.2.3}$$

where A is an $n \times n$ matrix and $y(x)$ and $f(x)$ are $n \times 1$ vectors, or in the form

$$\frac{d^n y}{dx^n} + a_{n-1}(x) \frac{d^{n-1} y}{dx^{n-1}} + \cdots + a_1(x) \frac{dy}{dx} + a_0(x)y = b(x). \tag{I.2.4}$$

We may, of course, put (I.2.4) in the form (I.2.3) by using the substitutions

$$\begin{aligned} u_1(x) &= y(x), \\ u_2(x) &= y'(x), \\ &\vdots \\ u_n(x) &= y^{n-1}(x). \end{aligned} \tag{I.2.5}$$

Then we have

$$\begin{aligned} u_1'(x) &= y'(x) = u_2(x) \\ u_2'(x) &= y''(x) = u_3(x) \\ &\vdots \\ u_n'(x) &= y^n(x) = -a_{n-1} u_n(x) - \cdots - a_0(x) u_1(x) + b(x). \end{aligned} \tag{I.2.6}$$

In matrix form this becomes

$$u'(x) = A(x)u(x) + b(x), \tag{I.2.7}$$

where

$$A(x) = \begin{pmatrix} 0 & 1 & 0 & \cdots & 0 \\ 0 & 0 & 1 & \cdots & 0 \\ & & \vdots & & \\ -a_0(x) & -a_1(x) & -a_2(x) & \cdots & -a_{n-1}(x) \end{pmatrix} \qquad b(x) = \begin{pmatrix} 0 \\ 0 \\ \vdots \\ b(x) \end{pmatrix}$$

We shall be particularly interested in the linear second order equation

$$\frac{d^2y}{dx^2} + a(x)\frac{dy}{dx} + b(x)y = c(x). \tag{I.2.8}$$

If $c(x) \equiv 0$, then (I.2.8) is said to be a *homogeneous* equation; otherwise, it is *nonhomogeneous*. The general solution of (I.2.8) involves two arbitrary constants. We normally write the general solution in the form

$$y(x) = y_h(x) + y_p(x), \tag{I.2.9}$$

where $y_h(x)$ is the general solution of the homogeneous equation and $y_p(x)$ is a particular solution of the nonhomogeneous equation. The general solution of the homogeneous equation is then written as

$$y_h(x) = Ay_1(x) + By_2(x), \tag{I.2.10}$$

where A and B are arbitrary constants and $y_1(x)$ and $y_2(x)$ are independent solutions of the homogeneous equation. (The concept of linearly independent solutions will be discussed in detail in section 3.) To see that $y_h(x)$ is also a solution of the homogeneous equation, we substitute (I.2.10) into (I.2.8) with $c(x)=0$. We get

$$Ay_1'' + By_2'' + a(x)(Ay_1' + By_2') + b(x)(Ay_1 + By_2) = 0,$$

or

$$(Ay_1 + By_2)'' + a(x)(Ay_1 + By_2)' + b(x)(Ay_1 + By_2) = 0.$$

Now, using (I.2.10), we obtain

$$y_h'' + a(x)y_h' + b(x)y_h = 0. \tag{I.2.11}$$

In order to evaluate A and B, we need two additional conditions. If both of these conditions are given at the same value of x, then the problem is said to be an *initial-value* problem. These conditions will normally take the form $y(0)=\alpha$, $y'(0)=\beta$. If the conditions are specified at different values of x, say $x=0$ and $x=1$, then the problem is said to be a *boundary-value* problem. In this case, we call it a *two-point* boundary-value problem.

Consider the equation

$$y' = ay, \quad y(0) = \alpha, \tag{I.2.12}$$

where a is a constant. In order to solve this problem, we first divide both sides by y and then integrate both sides from 0 to x. We obtain

$$\int_0^x \frac{1}{y} \frac{dy}{dx_1} dx_1 = a \int_0^x dx_1, \qquad \ln y(x) - \ln \alpha = ax$$

or

$$y(x) = \alpha e^{ax}. \qquad (\text{I.2.13})$$

We now consider the second order equation

$$y'' + ay' + by = 0, \qquad (\text{I.2.14})$$

where a and b are constants. Since the solution of (I.2.12) was an exponential, perhaps the solutions of (I.2.14) may also be exponential. Let $y(x) = e^{\lambda x}$. Substitution of $e^{\lambda x}$ into (I.2.14) yields

$$\lambda^2 e^{\lambda x} + a\lambda e^{\lambda x} + be^{\lambda x} = e^{\lambda x}(\lambda^2 + a\lambda + b) = 0.$$

Hence, (I.2.14) has solutions of the form $e^{\lambda x}$ if λ satisfies the *characteristic equation*

$$\lambda^2 + a\lambda + b = 0. \qquad (\text{I.2.15})$$

From elementary algebra we know that there are two roots of (I.2.15), and these roots must fall into one of three possibilities:
1. two real distinct roots, λ_1 and λ_2
2. two complex conjugate roots, λ_1 and λ_2
3. two real and equal roots.

In the first two cases we have $y_1(x) = e^{\lambda_1 x}$ and $y_2(x) = e^{\lambda_2 x}$. In the third case $\lambda_1 = \lambda_2 = \lambda$ implies $y_1(x) = y_2(x)$ and, hence, they are not independent. However, $y_1(x) = e^{\lambda x}$ and $y_2(x) = xe^{\lambda x}$ are independent. (See exercise 3.)

Definition: A collection $y_1, y_2, \ldots, y_n$, on $a < x < b$ of solutions of the n-dimensional first-order linear system,

$$y' = A(x)y, \qquad (\text{I.2.16})$$

is called a *fundamental system of solutions* of (I.2.16) if it is linearly independent.

We can generate a fundamental system by taking the solutions $y_1(x)$,

$y_2(x), ..., y_n(x)$ of (I.2.16) satisfying the initial condition $y_j(a) = e_j, j = 1, ..., n,$ where

$$e_j = (0, 0, ..., 1, 0, ... 0)^t, \quad 1 \text{ in the } j\text{th place}.$$

We introduce the $n \times n$ matrix $Y(x)$, whose jth column is $y_j(x)$, as introduced above. Then the matrix $Y(x)$, called the *fundamental matrix*, is the solution of the matrix differential equation

$$Y' = A(x)Y \tag{I.2.17}$$

satisfying $Y(a) = I$. Furthermore, the solution $y(x)$ of (I.2.16) satisfying $y(a) = y_0$ can be written

$$y(x) = Y(x)y_0. \tag{I.2.18}$$

So far, we have been indiscriminately referring to "the solution" of a differential equation. At this point we don't know whether a solution of a given differential equation even exists, much less whether it is unique. Thus, we now state a theorem giving sufficient conditions for the existence and uniqueness of the solution of an initial-valued system of first-order equations in the form

$$y'(x) = f[x, y(x)], \quad y(a) = y_0. \tag{I.2.19}$$

We shall always suppose that in the region $[a, b] X(-\infty, \infty)$ the function f satisfies

$$|f(x, z) - f(x, w)| \leqslant L|z - w|;$$

that is, the function f satisfies a *Lipschitz condition* with respect to its second variable. The constant L is called the Lipschitz constant.

Theorem 1: If $f(x, y)$ is continuous on $[a, b] X(-\infty, \infty)$ and satisfies the above Lipschitz condition on the region, then the problem

$$y'(x) = f[x, y(x)], \quad y(a) = \alpha, \tag{I.2.20}$$

has a unique solution on $[a, b]$ for every prescribed α.

However, the analysis is not so simple for boundary-value problems. For example, although the equation

$$y'' + \pi^2 y = 0, \tag{I.2.21}$$

subject to

$$y(0) = 0, \quad y'(0) = 1, \tag{I.2.22}$$

has a unique solution on any finite interval, the boundary-value problems defined by (I.2.21) and

$$y(0) = 0, \quad y(1) = 0, \tag{I.2.23}$$

or

$$y(0) = 0, \quad y(1) = 1, \tag{I.2.24}$$

or

$$y(0) = 0, \quad y'(1) = 1 \tag{I.2.25}$$

have, respectively, an infinite number of solutions, no solution, and a unique solution. Further discussion of boundary-value problems will be given in section 5.

For an example of the use of the Lipschitz condition, consider the general first-order linear problem

$$y'(z) = f(z) y(z) + g(z). \tag{I.2.26}$$

Then

$$|f(z, u) - f(z, v)| = |f(z)u + g(z) - f(z)v - g(z)|$$
$$= |f(z)| |u - v|.$$

If $f(z)$ and $g(z)$ are bounded on $[a, b]$ (i.e., $|f(z)| < L$ and $|g(z)| < M$), then f satisfies a Lipschitz condition with the Lipschitz constant L. Thus we can restate the uniqueness theorem for a linear problem to require only that the coefficients be continuous. Actually, this requirement, too, can be weakened somewhat.

3   Linearly Independent Solutions and Wronskians

Consider

$$L(y) = 0, \quad a \leqslant x \leqslant b, \tag{I.3.1}$$

where L is a second-order linear differential operator. The general solution is given by

$$y = c_1 y_1 + c_2 y_2, \tag{I.3.2}$$

provided that y_1 and y_2 are linearly independent solutions of (I.3.1). If

$$\frac{y_1(x)}{y_2(x)} = f(x) \neq \text{constant},$$

then y_1 and y_2 are linearly independent.

If y_1 and y_2 are not linearly independent, we may find nonzero constants c_1 and c_2 so that

$$c_1 y_1(x) + c_2 y_2(x) \equiv 0, \qquad a \leqslant x \leqslant b. \tag{I.3.3}$$

This relation holds identically on $[a, b]$. Hence, it follows that we may differentiate and obtain

$$c_1 y_1'(x) + c_2 y_2'(x) \equiv 0. \tag{I.3.4}$$

Viewing (I.3.3, I.3.4) as a system of two equations for the two unknowns c_1 and c_2, we form the determinant

$$W(y_1, y_2) = \begin{vmatrix} y_1(x) & y_2(x) \\ y_1'(x) & y_2'(x) \end{vmatrix} = y_1(x) y_2'(x) - y_1'(x) y_2(x), \tag{I.3.5}$$

which we shall refer to as the *Wronskian*.

The identical vanishing of the Wronskian is a *necessary* condition for the linear dependence of the functions $y_1(x)$ and $y_2(x)$, whereas the nonvanishing of the Wronskian is a *sufficient* condition for the linear independence of y_1 and y_2.

For equations with continuous coefficients, the nonvanishing of the Wronskian becomes both a *necessary* and a *sufficient* condition for the linear independence of y_1 and y_2. Consider the linear second-order problem

$$y'' + a_0 y' + a_1 y = 0, \tag{I.3.6}$$

where a_0 and a_1 are continuous functions on $[a, b]$. Now consider

$$W(x) = y_1(x) y_2'(x) - y_1'(x) y_2(x). \tag{I.3.7}$$

Differentiating, we obtain

$$\begin{aligned}
W'(x) &= y_1'(x) y_2'(x) + y_1(x) y_2''(x) - y_1''(x) y_2(x) - y_1'(x) y_2'(x) \\
&= y_1(x)[-a_0 y_2'(x) - a_1 y_2(x)] - y_2(x)[-a_0 y_1'(x) - a_1 y_1(x)], \\
&= -a_0[y_1(x) y_2'(x) - y_1'(x) y_2(x)] = -a_0 W(x).
\end{aligned}$$

Thus we see that the Wronskian satisfies the linear first-order equation

$$W'(x) = -a_0 W(x). \tag{I.3.8}$$

The solution of (I.3.8) is

$$W(x) = W(x_0) e^{-\int_{x_0}^{x} a_0(z)\, dz}, \qquad x_0 \in [a, b]. \tag{I.3.9}$$

If $W(x_0)=0$, then $W(x)\equiv 0$. Thus, if the Wronskian $W(x_0)\neq 0$, then y_1 and y_2 are independent.

The above result enables us to generate linearly independent solutions easily. We need only let y_1 be a solution of (I.3.6), satisfying the initial conditions

$$y_1(0) = 1, \qquad y_1'(0) = 0,$$

and y_2 be a solution of (I.3.6), satisfying the initial conditions

$$y_1(0) = 0, \qquad y_2'(0) = 1.$$

The Wronskian at $x=0$ for such a system is

$$W(0) = y_1(0)\, y_2'(0) - y_1'(0)\, y_2(0) = 1.$$

4   The Sturmian Theory

In this section, we shall discuss some of the basic concepts of the oscillation and comparison theory for second-order equations attributable to Sturm. This material is very basic to the study of eigenvalue problems discussed in detail in Chapter V. We shall consider equations of the form

$$\frac{d}{dx}\left\{K \frac{dy}{dx}\right\} - Gy = 0, \tag{I.4.1}$$

where the function G is continuous on the interval $[a, b]$ and the function K is assumed to be strictly positive and continuously differentiable on $[a, b]$.

Theorem 2: No continuous solution, except the identically zero solution, can have an infinite number of zeros in $[a, b]$.

Proof: Suppose there is such a solution. By the Bolzano-Weierstrass theorem, we would have at least one limit point of these zeros. Denote this limit point by c. Consider

$$y(c + h) = y(c) + hy'(c + \xi h), \quad 0 \leqslant \xi \leqslant 1. \tag{I.4.2}$$

Since c is a limit point of the zeros, we can find an h so small that

$$y(c + h) = 0. \tag{I.4.3}$$

Then, from (I.4.2), we have

$$y'(c + \xi h) = 0. \tag{I.4.4}$$

Since $y(x)$ is assumed to be continuous, it follows that

$$y'(c) = 0. \tag{I.4.5}$$

However, (I.4.1) subject to the conditions $y(c)=0$ and $y'(c)=0$ has only the trivial solution. Hence, there is no solution with an infinite number of zeros except the trivial solution.

Theorem 3 (The Separation Theorem): The zeros of two real linearly independent solutions of a linear differential equation of the second order separate one another.

Proof: Let y_1 and y_2 be any two real linearly independent solutions of the given equation. Assume that (a, b) is large enough that y_1 vanishes at least twice in that interval. Denote any two consecutive zeros of y_1 in that interval by x_1 and x_2. We wish to prove that y_2 vanishes at least once in the open interval $x_1 < x < x_2$.

First of all, y_2 cannot vanish at x_1 or x_2, for y_2 would then be a multiple of y_1. Suppose y_2 does not vanish in (x_1, x_2). Consider the function y_1/y_2. It is easy to see that it is continuously differentiable in $[x_1, x_2]$ and vanishes at the endpoints. Hence, by Rolle's theorem, its derivative, given by

$$\frac{d}{dx}(y_1/y_2) = \frac{y_2 y_1' - y_1 y_2'}{y_2^2}, \tag{I.4.6}$$

vanishes at least once.

Comparing (I.4.6) with (I.3.5), we see that the numerator is merely the Wronskian and, hence, cannot vanish in (x_1, x_2). Our original assumption is contradicted and, thus, y_2 must vanish at least once in (x_1, x_2). We must also show that it cannot vanish more than once. This is easy to prove. If it were shown to vanish more than once, by reversing the roles of y_1 and y_2 in the above argument, y_1 would have to vanish between them, thereby contradicting that x_1 and x_2 are consecutive zeros of y_1.

For example, the zeros of $\cos x$ and $\sin x$ separate one another, since they are linearly independent solutions of $y'' + y = 0$.

If a solution has no more than one zero on any given interval, it is said to be nonoscillatory in that interval. If two functions $y_1(x)$ and $y_2(x)$ are continuous on the interval (a, b) and if $y_1(x)$ has more zeros in the interval than does $y_2(x)$, then $y_1(x)$ is said to oscillate more rapidly than $y_2(x)$.

Theorem 4 (Sturm's Fundamental Theorem): If the solutions of

$$\frac{d}{dx}\left\{ K \frac{dy}{dx} \right\} - Gy = 0 \tag{I.4.7}$$

oscillate in the interval $[a, b]$, they will oscillate more rapidly when K and G are diminished.

Proof: We shall prove the theorem only for K unchanged and G diminished and leave the proof for both being diminished to exercise 7. Let y_1 be a non-trivial solution of

$$\frac{d}{dx}\left\{ K \frac{dy_1}{dx} \right\} - G_1 y_1 = 0 \tag{I.4.8}$$

and y_2 be a nontrivial solution of

$$\frac{d}{dx}\left\{ K \frac{dy_2}{dx} \right\} - G_2 y_2 = 0, \tag{I.4.9}$$

where G_1 is pointwise greater than G_2 on $[a, b]$. If we multiply (I.4.8) by $y_2(x)$ and (I.4.9) by $y_1(x)$, subtract, and then integrate both sides from x_1 to x_2, we get

$$K\left[y_1'(x) y_2(x) - y_1(x) y_2'(x) \right]\big|_{x_1}^{x_2} = \int_{x_1}^{x_2} (G_1 - G_2) y_1 y_2 \, dx. \tag{I.4.10}$$

Let the limits of integration x_1 and x_2 be consecutive zeros of $y_1(x)$ and assume that $y_2(x)$ has no zero in (x_1, x_2). Then y_1 and y_2 may be assumed to be positive, and we conclude that the right hand side of (I.4.10) is positive. Since y_1 and y_2 are assumed to be positive on the interval and y_1 is zero at the endpoints, y_1' is positive at x_1 and negative at x_2. This implies that the left hand side of (I.4.10) is negative. Hence, y_2 must vanish at least once between x_1 and x_2.

An important consequence of the theorem is that if y_1 and y_2 are both zero at x_1, then y_2 has a zero $x_3 < x_2$. That is, y_2 oscillates more rapidly than y_1.

Not all solutions, of course, oscillate. In many applications we want to know (1) whether a given equation possesses oscillatory solutions and (2) the distance between zeros of the solutions which do oscillate. We seek the answers by developing comparison equations which have constant coefficients.

Since the functions $K(x)$ and $G(x)$ are continuous on the closed interval $[a, b]$, it follows that both K and G are bounded. Let the upper and lower bounds of K and G be given, respectively, by K, G and $k > 0$, g. Replacing K and G in (I.4.7) by their upper and lower bounds, we obtain the two comparison equations

$$\frac{d^2y}{dx^2} - \frac{g}{k} y = 0 \tag{I.4.11}$$

and

$$\frac{d^2y}{dx^2} - \frac{G}{K} y = 0. \tag{I.4.12}$$

According to Theorem 4, the solutions of (I.4.7) do not oscillate more rapidly than those of (I.4.11), and the solutions of (I.4.7) oscillate at least as rapidly as those of (I.4.12). The solutions of (I.4.11) can be divided into two parts:

(1) If $g > 0$, the solutions are exponential and, hence, have no zeros in $[a, b]$. If $g = 0$, the solution is linear and, hence, nonoscillatory for $g \geqslant 0$. It then follows that *if $G \geqslant 0$ throughout the interval $[a, b]$, the solutions of (I.4.7) are nonoscillatory.*

(2) If $g < 0$, there is the oscillatory solution $\sin \sqrt{-g/k}\, x$; the interval between its consecutive zeros is

$$\pi \sqrt{-k/g}.$$

Hence, if $\pi \sqrt{-k/g} > b - a$, no solution of (I.4.7) can have more than one zero in (a, b). That is, the solutions of (I.4.7) are nonoscillatory for

$$-g/k < \frac{\pi^2}{(b-a)^2}.$$

If $G < 0$, the solutions of (I.4.12) are oscillatory and the interval between zeros is $\pi \sqrt{-K/G}$. Thus, a sufficient condition for the solutions of (I.4.7) to oscillate is that

$$- G/K \geqslant \frac{\pi^2}{(b-a)^2}.$$

5   Boundary-Value Problems

As we have seen in section 2, a very naive-appearing linear boundary-value problem such as

$$y'' + \pi^2 y = 0 \tag{I.5.1}$$

can have a unique solution, no solution, or an infinite number of solutions, depending upon the type of boundary conditions chosen. A thorough examination of the existence and uniqueness of solutions of boundary-value problems is beyond the scope of this book. We shall present only a few of the basic ideas; but enough, hopefully, that the reader will be aware of some of the pitfalls.

Consider the linear boundary-value problem

$$y''(z) + f(z) y'(z) + g(z) y(z) = 0, \tag{I.5.2a}$$

$$y(0) = \alpha, \quad y(x) = \beta. \tag{I.5.2b}$$

We seek the solution of (I.5.2) on the interval $a \leqslant z \leqslant x$. The existence and uniqueness of solutions of (I.5.2a) subject to initial conditions have been established in section 2. For example, we need only require $f(z)$ and $g(z)$ to be continuous on $[0, x]$. Hence, we shall study the solution of the boundary value problem by studying the solution of the associated initial-value problem

$$w''(z) + f(z) w'(z) + g(z) w(z) = 0, \tag{I.5.3a}$$

$$w(0) = \alpha, \quad w'(0) = \gamma. \tag{I.5.3b}$$

Recall that every solution of (I.5.2a) is a linear combination of two linearly independent solutions, $y_1(z)$ and $y_2(z)$, of (I.5.2a) which satisfy, say,

$$y_1(0) = 1, \quad y_1'(0) = 0, \tag{I.5.4}$$

$$y_2(0) = 0, \quad y_2'(0) = 1. \tag{I.5.5}$$

Then the unique solution of (I.5.3) may be written as

$$w(z) = \alpha y_1(z) + \gamma y_2(z). \tag{I.5.6}$$

If we choose γ in (I.5.3b) so that

$$w(x) = \beta = \alpha y_1(x) + \gamma y_2(x) \qquad (\text{I.5.7})$$

or

$$\gamma = \frac{\beta - \alpha y_1(x)}{y_2(x)}, \qquad (\text{I.5.8})$$

then $y(z) \equiv w(z)$. The value of γ is uniquely determined, provided that $y_2(x) \neq 0$. In general, if $y_2(x) = 0$, there is no solution to the nonhomogeneous problem (I.5.2). A solution would exist if $\beta = \alpha y_1(x)$, but it would not be unique since $w(z)$ would be a solution of (I.5.2) for arbitrary γ.

If $y_2(x) = 0$, then $y_2(z)$ satisfies (I.5.2a) with the boundary conditions

$$y_2(0) = 0, \quad y_2(x) = 0. \qquad (\text{I.5.9})$$

Since $y_2'(0) = 1$, there exists a nontrivial solution of the homogeneous problem (I.5.2a), (I.5.9).

The analysis can be easily extended to the nonhomogeneous equation

$$y''(z) + f(z) y'(z) + g(z) y(z) = r(z). \qquad (\text{I.5.10})$$

We need only define a new function by the relation

$$u(z) \equiv y(z) - y_p(z), \qquad (\text{I.5.11})$$

where $y_p(z)$ is a particular solution of (1.5.10). The function $u(z)$ then satisfies the homogeneous equation (I.5.2a) subject to the boundary conditions

$$u(0) = y(0) - y_p(0) = \alpha', \qquad (\text{I.5.12a})$$

$$u(x) = y(x) - y_p(x) = \beta'. \qquad (\text{I.5.12b})$$

These results may be summarized in a theorem which is referred to as the *alternative principle*.

Theorem 5: Let $f(z)$, $g(z)$, and $r(z)$ be continuous on $[0, x]$. Then, for any α and β, the following mutually exclusive alternatives hold: Either the non-homogeneous boundary-value problem

$$Ly \equiv y''(z) + f(z) y'(z) + g(z) y(z) = r(z), \qquad (\text{I.5.13a})$$

$$y(0) = \alpha, \quad y(x) = \beta, \qquad (\text{I.5.13b})$$

has a unique solution or the homogeneous boundary-value problem,

$$Ly = 0 \tag{I.5.14a}$$

$$y(0) = 0, \quad y(x) = 0, \tag{I.5.14b}$$

has a nontrivial solution.

This theorem may be restated more simply as follows: *The problem* (I.5.13) *has a unique solution if and only if the problem* (I.5.14) *has only the trivial solution.* For nonlinear problems the disussion is more complicated. Consider

$$y'' = f(z, y, y'), \quad 0 < z < x, \tag{I.5.15a}$$

$$y(0) = \alpha, \quad y(x) = \beta. \tag{I.5.15b,c}$$

We shall approach this problem very much as we did the linear problem. That is, we shall study the associated initial-value problem

$$w'' = f(z, w, w'), \quad 0 \leqslant z \leqslant x, \tag{I.5.16a}$$

$$w(0) = \alpha, \quad w'(0) = \gamma. \tag{I.5.16b,c}$$

Denote the solution of (I.5.16) by

$$w = w(z; \gamma) \tag{I.5.17}$$

to emphasize the dependence of the solution upon the initial condition γ. We now evaluate the solution at $z = x$ and seek a value of γ so that

$$w(x; \gamma) = \beta. \tag{I.5.18}$$

For a linear problem there would be, at most, one solution to (I.5.18); however, for nonlinear problems (I.5.18) is generally a transcendental equation which may have several solutions. This leads us to the theorem.

Theorem 6: Let the function $f(z, u, v)$ be continuous on

$$R: 0 \leqslant z \leqslant x, \quad u^2 + v^2 < \infty,$$

and satisfy a uniform Lipschitz condition in u and v. Then the nonlinear boundary-value problem (I.5.15) has as many solutions as there are distinct roots, $\gamma = \gamma^{(i)}$ of (I.5.18). The solutions of (I.5.15) (i.e., the solutions of the initial-value problem (I.5.16) with initial condition $\gamma = \gamma^{(i)}$) are

$$y(z) = y^{(i)}(z) \equiv w(z, \gamma^{(i)}). \tag{I.5.19}$$

6   Green's Functions

Green's functions play a fundamental role in the study of boundary-value problems. In this section we shall discuss a method for constructing Green's functions. The method, however, is often undesirable for numerical computation. A recommended technique for numerical computation is discussed in section 8 of Chapter IV.

Let us consider the differential operator

$$Lu = \frac{d}{dz}\left(p(z)\frac{du}{dz}\right) + q(z)u, \qquad (I.6.1)$$

where $p'(z)$ and $q(z)$ are continuous on $a \leqslant z \leqslant b$ and $p(z) \neq 0$ on $a \leqslant z \leqslant b$. The boundary conditions will be of the form

$$\alpha_1 u(a) + \beta_1 u'(a) = 0 \qquad (I.6.2a)$$

$$\alpha_2 u(b) + \beta_2 u'(b) = 0. \qquad (I.6.2b)$$

Our primary purpose will be to solve the corresponding inhomogeneous equation

$$Lu = f(z), \qquad (I.6.3)$$

subject to the boundary conditions (I.6.2). We shall try to manufacture a solution of (I.6.2), (I.6.3) by the method of variation of parameters. Let $u_1(z)$ and $u_2(z)$ be linearly independent solutions of the homogeneous equation; that is, solutions of (I.6.3) with $f(z) \equiv 0$. In order to reduce some computations involved, we shall assume $u_1(z)$ to be a solution of

$$Lu_1 = 0, \qquad (I.6.4a)$$

subject to

$$\alpha_1 u_1(b) + \beta_1 u_1'(b) = 0, \qquad (I.6.4b)$$

and $u_2(z)$ to be a solution of

$$Lu_2 = 0, \qquad (I.6.5a)$$

subject to

$$\alpha_2 u_2(a) + \beta_2 u_2'(a) = 0. \qquad (I.6.5b)$$

How do we know that the $u_1(z)$ and $u_2(z)$ generated in this manner will be linearly independent? (See exercise 8.)

Write, as a try, the solution of (I.6.2), (I.6.3) as

$$u(z) = c_1(z) u_1(z) + c_2(z) u_2(z). \tag{I.6.6}$$

Since the two functions $c_1(z)$ and $c_2(z)$ are to be determined, we expect that two conditions ought to be imposed. One condition requires that the assumed solution satisfy the differential equation. The other condition may be imposed arbitrarily. Let us proceed. By differentiation of (I.6.6) we have

$$u'(z) = c_1(z) u_1'(z) + c_2(z) u_2'(z) + c_1'(z) u_1(z) + c_2'(z) u_2(z). \tag{I.6.7}$$

Further differentiation would continue to introduce more terms. Thus, we impose our arbitrary condition and require that

$$c_1'(z) u_1(z) + c_2'(z) u_2(z) \equiv 0. \tag{I.6.8}$$

Hence, (I.6.7) becomes

$$u'(z) = c_1(z) u_1'(z) + c_2(z) u_2'(z). \tag{I.6.9}$$

Multiplying both sides of (I.6.9) by $p(z)$, differentiating with respect to z, and adding $q(z) u(z)$ to both sides, we get

$$Lu = (pu')' + qu = c_1 (pu_1')' + c_2 (pu_2')' + c_1' pu_1'$$
$$+ c_2' pu_2' + qc_1 u_1 + qc_2 u_2 = f. \tag{I.6.10}$$

Since u_1 and u_2 both satisfy the homogeneous equation, (I.6.10) can be written as

$$c_1'(z) u_1'(z) + c_1'(z) u_2'(z) = f(z)/p(z). \tag{I.6.11}$$

Viewing (I.6.8) and (I.6.11) as a system of two equations in two unknowns, we can write the solutions for $c_1'(z)$ and $c_2'(z)$ as

$$c_1'(z) = \frac{- u_2(z) f(z)}{p(z) W(z)} \tag{I.6.12}$$

and

$$c_1'(z) = \frac{u_1(z) f(z)}{p(z) W(z)}, \tag{I.6.13}$$

where $W(z) = u_1(z) u_2'(z) - u_1'(z) u_2(z)$, the Wronskian. The functions $c_1'(z)$ and $c_2'(z)$ are well-defined, since $p(z) \neq 0$ and $W(z) \neq 0$.

Integrating (I.6.12) from a to z and (I.6.13) from b to z, we get

$$c_1(z) = -\int_a^z \frac{u_2(t)f(t)}{p(t)W(t)} dt + \hat{c}_1 \tag{I.6.14}$$

and

$$c_2(z) = -\int_z^b \frac{u_1(t)f(t)}{p(t)W(t)} dt + \hat{c}_2. \tag{I.6.15}$$

Then our solution (I.6.6) can be represented as

$$u(z) = -\int_a^z \frac{u_1(z)u_2(t)f(t)}{p(t)W(t)} dt - \int_z^b \frac{u_2(z)u_1(t)f(t)}{p(t)W(t)} dt + \hat{c}_1 u_1(z) + \hat{c}_2 u_2(z). \tag{I.6.16}$$

It can be easily shown (see exercises 12 and 13) that $p(z)\,W(z)=k$, a constant in $a \leqslant z \leqslant b$. The constants $\hat{c}_1$ and $\hat{c}_2$ must be evaluated from the boundary conditions

$$\alpha_1 u(a) + \beta_1 u'(a) = 0, \tag{I.6.17a}$$
$$\alpha_2 u(b) + \beta_2 u'(b) = 0. \tag{I.6.17b}$$

If we can evaluate (I.6.16) at $z=a$, we obtain

$$u(a) = -\frac{1}{k}\int_a^b u_2(a)u_1(t)f(t)\, dt + \hat{c}_1 u_1(a) + \hat{c}_2 u_2(a). \tag{I.6.18}$$

Differentiating in (I.6.16) with respect to z and then evaluating the result at $z=a$, we get

$$u'(a) = -\frac{1}{k}\int_a^b u_2'(a)u_1(t)f(t)\, dt + \hat{c}_1 u_1'(a) + \hat{c}_2 u_2'(a), \tag{I.6.19}$$

Multiplying (I.6.18) by α_1 and (I.6.19) by β_1 and adding, we have

$$\alpha_1 u(a) + \beta_1 u'(a) = -\frac{1}{k}\int_a^b u_1(t)f(t)\left[\alpha_1 u_2(a) + \beta_1 u_2'(a)\right] dt$$

$$+ \hat{c}_1\left[\alpha_1 u_1(a) + \beta_1 u_1'(a)\right] + \hat{c}_2\left[\alpha_1 u_2(a) + \beta_1 u_2'(a)\right]$$
$$= \hat{c}_1\left[\alpha_1 u_1(a) + \beta_1 u_1'(a)\right]. \tag{I.6.20}$$

Since $\alpha_1 u(a) + \beta_1 u'(a) = 0$, we must take $\hat{c}_1 = 0$. (Why not take $\alpha_1 u_1(a) + \beta_1 u'(a) = = 0$?). If we consider the point $z = b$, we can set $\hat{c}_2 = 0$ and satisfy (I.6.17b), and, of course,

$$Lu = f. \tag{I.6.21}$$

(The above analysis was simplified by the choice of the conditions, (I.6.4b) and (I.6.5b). However, any other conditions where $u_1(z)$ and $u_2(z)$ are linearly independent would have produced the desired results.)

Hence, the solution is given by

$$u(z) = -\frac{1}{k} \int_a^z u_1(z) u_2(t) f(t) dt - \frac{1}{k} \int_z^b u_2(z) u_1(t) f(t) dt.$$

$$\tag{I.6.22}$$

Now define the function $G(z, t)$ by

$$G(z, t) = \begin{cases} -\dfrac{1}{k} u_1(z) u_2(t), & t \leqslant z, \\[2mm] -\dfrac{1}{k} u_1(t) u_2(z), & z \leqslant t, \end{cases} \tag{I.6.23}$$

where $k = p(z_0) W(z_0)$ for any $z_0 \in [a, b]$. Then the solution of (I.6.3) can be written as

$$u(z) = \int_a^b G(z, t) f(t) dt. \tag{I.6.24}$$

The function $G(z, t)$, called the *Green's Function* for the operator L and the boundary conditions (I.6.17), satisfies the following conditions:*

1. $G(z, t)$ is continuous in z and t, $a \leqslant z, t \leqslant b$.
2. $G(z, t)$ satisfies the boundary condition as a function of z and as a function of t.
3. $G(z, t)$, considered as a function of z, satisfies the differential equation (I.6.1) at all points of the interval (a, b) except at $z = t$.
4. The derivative $G_z(z, t)$ is continuous only for $a \leqslant z < t$ and $t < z \leqslant b$, and has a

---

* The Green's Function not only satisfies these five conditions but it can also be determined from these conditions.

jump discontinuity of $-1/p(t)$ at $z=t$; that is,

$$G_z(t+0, t) - G_z(t-0, t) = -1/p(t).$$

5. $G(z, t) = G(t, z)$; that is, G is symmetric.

7 Riccati Equations – Scalar Case

One of the most famous of all differential equations is the Riccati equation, named after Jacopo Francesco, Count Riccati (1676–1754), who studied what is equivalent to the following form in 1724:

$$y'(z) + ay^2(z) = bz^n. \tag{I.7.1}$$

The general form of the Riccati equation is written as

$$y'(z) = a(z) + b(z) y(z) + c(z) y^2(z). \tag{I.7.2}$$

This equation is, of course, first-order and nonlinear. As a consequence of its being first-order, the general solution involves only one arbitrary constant. Hence, only one additional condition need be given to specify its solution

$$y(z_0) = y_0. \tag{I.7.3}$$

The analysis of the Riccati equation is complicated for two reasons: Its solution, in general, cannot be obtained in closed form and its solution may not exist for all $z \geqslant z_0$, even though the coefficients may be very smooth. For example, we shall encounter such a problem in section 8 of Chapter III. There we will have $a = \sigma/c$, $b = 0$, and $c = \sigma c$. The solution is

$$y(z) = \frac{1}{c} \tan \sigma z \tag{I.7.4}$$

which has maximal intervals of existence of length π/σ.

It is well-known that the transformation

$$w(z) = e^{-\int_{z_0}^{z} c(s) \, y(s) \, ds} \tag{I.7.5}$$

plays a fundamental role in the study of Riccati equations. By differentiating in (I.7.5), we find that

$$w'(z) = -c(z)\,y(z)\,w(z). \tag{I.7.6}$$

Differentiating again and using (I.7.2), we can write

$$w''(z) = -[c(z)\,y(z)]'\,w(z) - c(z)\,y(z)\,w'(z). \tag{I.7.7}$$

Further algebraic manipulation, using (I.7.2) and (I.7.6), yields

$$w''(z) + \left[b(z) - \frac{c'(z)}{c(z)}\right]w'(z) + a(z)\,c(z)\,w(z) = 0. \tag{I.7.8}$$

This equation is a second-order linear differential equation and illustrates the close connection between linear second-order problems and the Riccati equation. Equations (I.7.5) and (I.7.6) are commonly referred to as *Riccati transformations*.

Actually, our primary interest in the connection between Riccati equations and linear second-order equations will be to derive a Riccati equation from a given second-order equation. Consider a second-order equation of the form

$$u''(z) + m(z)\,u'(z) + n(z)\,u(z) = 0 \tag{I.7.9}$$

and define the function $r(z)$ by the relation

$$u(z) = r(z)\,u'(z). \tag{I.7.10}$$

If we differentiate in (I.7.10) and use (I.7.9) and then (I.7.10) again, we obtain

$$u'(z)\,\{1 + m(z)\,r(z) + n(z)\,r^2(z) - r'(z)\} = 0. \tag{I.7.11}$$

We argue that, since (I.7.11) must hold for all z, the term in braces must be zero. Thus,

$$r'(z) = 1 + m(z)\,r(z) + n(z)\,r^2(z), \tag{I.7.12}$$

a Riccati equation.

It is sometimes convenient to define another function $s(z)$ by the relation

$$u'(z) = s(z)\,u(z). \tag{I.7.13}$$

Using the same reasoning as before, we see that $s(z)$ satisfies

$$-s'(z) = n(z) + m(z)\,s(z) + s^2(z). \tag{I.7.14}$$

It is evident on a common domain of definition that

$$r(z) = 1/s(z). \tag{I.7.15}$$

This should not be too surprising, since any nonsingular transformation of the form

$$r(z) = \frac{\alpha s(z) + \beta}{\gamma s(z) + \delta} \tag{I.7.16}$$

converts a Riccati equation into another Riccati equation (see exercise 19). The coefficients α, β, γ, δ need not be constants.

Notice that if $u(z) = 0$, then $r(z)$ as defined by (I.7.10) becomes ill-behaved. From the fundamental uniqueness theorem, $u(z)$ and $u'(z)$ cannot both be zero at the same point. Thus, when $r(z)$ is ill-behaved, $s(z)$ as defined by (I.7.13) is well-defined. These transformations play a fundamental role in the study of eigenvalue problems to be discussed in Chapter V.

Another transformation, which is a special case of (I.7.16), that is important in the study of Riccati equations is

$$y(z) = K(z) t(z) + L(z), \tag{I.7.17}$$

where $K(z)$ and $L(z)$ are differentiable. Substituting this transformation into (I.7.2), we get

$$\begin{aligned} K(z) t'(z) = \quad &\{a(z) + b(z) L(z) + c(z) L^2(z) - L'(z)\} \\ &+ \{b(z) K(z) + 2c(z) K(z) L(z) - K'(z)\} t(z) \\ &+ c(z) K^2(z) t^2(z). \end{aligned} \tag{I.7.18}$$

The form of (I.7.18) suggests setting

$$\alpha K(z) = b(z) K(z) + 2c(z) K(z) L(z) - K'(z), \tag{I.7.19}$$

$$\beta K(z) = c(z) K^2(z), \tag{I.7.20}$$

where α and β are arbitrary constants with $\beta \neq 0$. Explicitly, this amounts to choosing

$$K(z) = \beta/c(z), \tag{I.7.21}$$

$$L(z) = \frac{1}{2c(z)} \left\{ \alpha - b(z) + \frac{K'(z)}{K(z)} \right\}. \tag{I.7.22}$$

Dividing (I.7.18) by $K(z)$ and setting

$$f(z) = \frac{1}{K(z)} \{a(z) + b(z) L(z) + c(z) L^2(z) - L'(z)\},$$

we obtain the Riccati equation

$$t'(z) = f(z) + \alpha t(z) + \beta t^2(z). \tag{I.7.24}$$

The above manipulation requires that $c(z) \neq 0$. This particular form of the Riccati equation is particularly useful, since α and β are constants of our choice except that $\beta \neq 0$. We might, for example, choose $\alpha = 0$ and $\beta = 1$.

8   Riccati Equations – Matrix Case

The form of the matrix Riccati equation with which we shall be primarily concerned is given by

$$R'(z) = B(z) + A(z) R(z) + R(z) D(z) + R(z) C(z) R(z), \tag{I.8.1}$$

where all matrices are $n \times n$ continuous functions of z.

As in the scalar case, we can relate the above Riccati equation to a system of linear differential equations. Consider the differential equations

$$U'(z) = A(z) U(z) + B(z) V(z), \tag{I.8.2a}$$

$$-V'(z) = C(z) U(z) + D(z) V(z), \tag{I.8.2b}$$

subject to the initial conditions

$$U(0) = 0, \tag{I.8.3a}$$

$$V(0) = I, \tag{I.8.3b}$$

where U and V are $n \times n$ matrices and I is the $n \times n$ identity matrix.

If the initial condition on R is given by

$$R(0) = 0, \tag{I.8.4}$$

then the relationship between R and the above system is given by

$$U(z) = R(z) V(z). \tag{I.8.5}$$

In order to show this we differentiate in (I.8.5) with respect to z and obtain

$$U'(z) = R'(z) V(z) + R(z) V'(z). \tag{I.8.6}$$

(Since we are dealing with matrices the order of operation must be carefully observed.)

Using (I.8.2a, b) and (I.8.5), we have

$$\{R'(z) - B(z) - A(z) R(z) - R(z) D(z) - R(z) C(z) R(z)\} V(z) = 0. \tag{I.8.7}$$

Since the coefficients A, B, C and D are continuous, $V(z)$ is nonsingular, at least in some neighborhood of the origin. Thus,

$$R'(z) = B(z) + A(z) R(z) + R(z) D(z) + R(z) C(z) R(z). \tag{I.8.8}$$

Evaluating (I.8.5) at $z=0$, we see that the initial conditions are satisfied.

9   Exercises

1. Consider the linear differential equation $y' + ay = f(t)$, $y(0) = c$, and write the solution in the form $y = L(t, c, f)$, indicating explicitly the dependence upon t, c and the forcing term $f(t)$. Show that

   (a) $L(t, c, f) = L_1(t, c) + L_2(t, f)$,   where
   (b) $L_1(t, c_1 + c_2) = L_1(t, c_1) + L_2(t, c_2)$
   (c) $L_2(t, f_1 + f_2) = L_2(t, f_1) + L_2(t, f_2)$.

   (Bellman and Kalaba)

2. Consider the equation $y''(z) + a(z) y'(z) + b(z) y(z) = c(z)$, whose solution for a particular set of initial or boundary conditions is unique on the interval $0 \leqslant z \leqslant T$. Let x_0 and x_1 be any two points where $0 \leqslant x_0 < x_1 \leqslant T$. Show that any one of the four values $y(x_0)$, $y'(x_0)$, $y(x_1)$, $y'(x_1)$ can be written as a linear combination of any two of the remaining three. For example: $y(x_0) = \alpha y'(x_0) + \beta y(x_1) + \gamma$.
   (Hint: Write $y(z)$ in the form $y(z) = Af(z) + Bg(z) + h(z)$, where $Af(z) + Bg(z)$ is the general solution of the homogeneous equation, and $h(z)$ is a particular solution of the inhomogeneous equation. Evaluate $y(z)$ and

$y'(z)$ at x_0 and solve for A and B. Substitute the results into the general solution for $y(z)$ and again evaluate at x_1. Then use Cramer's rule to solve for desired quantities.)

3. When the two roots of the characteristic equation (I.2.15) are real and equal, we have only one solution, $e^{\lambda x}$. In order to get a second linearly independent solution, we use the method of reduction of order. Let

$$y = v(x) e^{\lambda x}.$$

Substitute this solution into (I.2.14) and show that

$$v(x) = c_1 x + c_2,$$

which is, therefore, the general solution of (I.2.14) when $\lambda_1 = \lambda_2$ is given by

$$y(x) = c_1 e^{\lambda x} + c_2 x e^{\lambda x}.$$

4. Let y_1 be a solution of $y' + ay = g(x)$, $y(0) = c_1$, and y_2 the solution with $y_2(0) = c_2$. If $c_1 \neq c_2$, can $y_1(x) = y_2(x)$ for any $x > 0$?

(Bellman)

5. Solve completely: $y' = f(x, y)$, $y(0) = 0$, where $f(x, 0) \equiv 0$ and f satisfies the Lipschitz condition.

(Burgmeier)

6. Solve $y'' + y = 0$ with $y(0) = 0$, $y(x) = 1$.
   What restrictions, if any, must be imposed on x in order to insure that the above problem has a unique solution?

7. Prove theorem 4 of section 4 when both K and G are diminished.

8. Prove that $u_1(z)$ and $u_2(z)$ of section 6 are linearly independent.

9. Prove that the Green's function satisfies conditions 1–5 given at the end of section 6.

10. Derive the Green's function for

$$Lu = u'' + u,$$
$$u(0) = 0, \quad u'(x) = 0.$$

11. Derive the Green's function for

$$Lu = u'' - u,$$
$$u(0) = 0, \quad u'(x) = 0.$$

12. Let u_1 and u_2 be independent solutions of

(a)
$$Lu = \frac{d}{dz}\left[p(z)\frac{du}{dz}\right] + q(z)u(z) = 0.$$

Show that

(b)
$$u_1 Lu_2 - u_2 Lu_1 = u_1 \frac{d}{dz}\left[p(z)\frac{du_2}{dz}\right] - u_2 \frac{d}{dz}\left[p(z)\frac{dz}{dz}\right] = 0.$$

13. Integrate (b) from z_1 to z_2, where z_1 and z_2 are any two points belonging to $a \leqslant z \leqslant b$. Integrate the result by parts and show that $p(z)\,W(z)$ is a constant for all z in $a \leqslant z \leqslant b$.

14. By considering (1.6.16) at $z=b$, show that $c_2=0$.

15. Use a handbook of tables to solve

$$r' = a + br + cr^2, \, r(0) = 0, \quad a, b, c, \text{ nonzero constants}.$$

Graph the behavior of the solutions.

16. Without solving, discuss and graph the behavior of the solutions of the Riccati equation

$$r' = -ar + br^2, \quad \text{for} \quad -\infty < r(0) < \infty,$$

where a and b are positive constants.

17. Without solving, discuss and graph the behavior of the solutions of the Riccati equation

$$-s' = -as + bs^2, \quad \text{for} \quad -\infty < s(0) < \infty,$$

where a and b are positive constants.

18. If $r' = (1-r)^2/\varepsilon$, $r(0)=0$, where ε is a constant, show that

$$1 - e^{x/\varepsilon} \leqslant r \leqslant 1 - e^{-2x/\varepsilon}$$

(Nelson)

19. Show that, if $r(z)$ satisfies a Riccati equation of the form

$$r'(z) = a(z) + b(z) r(z) + c(z) r^2(z),$$

then $s(z) = (\alpha r(z) + \beta)/(\gamma r(z) + \delta)$, where α, β, γ, δ are constants, and $\alpha\delta - \beta\gamma \neq 0$ also satisfies a Riccati equation of the same form.

20. Find K and L so that the transformation $r = Kt + L$ converts the Riccati equation

$$r' = 1 + zr^2$$

into a Riccati equation of the form

$$t' = f(z) + t^2.$$

21. Show that if $R(z)$ satisfies (I.8.1), then $S(z) = R^{-1}(z)$ satisfies the Riccati equation

$$- S'(z) = C(z) + S(z) A(z) + D(z) S(z) + S(z) B(z) S(z).$$

22. Show that, if $R(z)$ satisfies a Riccati equation, then

$$S(z) = (KR(z) + L)^{-1} (MR(z) + N)$$

satisfies a Riccati equation, provided that $KN - ML \neq$ singular matrix.

10   Bibliographical Discussion

Sections 1–3

There are a number of good books on the subject of ordinary differential equations. A few of these, in order of increasing sophistication, are

W. E. Boyce and R. C. DiPrima, *Introduction to Differential Equations*, John Wiley and Sons, Inc., New York, 1970.

G. Birkhoff and G.-C. Rota, *Ordinary Differential Equations*, Ginn and Co., Boston, 1962.

E. L. Ince, *Ordinary Differential Equations*, Dover Publ., Inc., New York, 1944.

E. A. Coddington and N. Levinson, *Theory of Ordinary Differential Equations*, McGraw-Hill Book Co. Inc., New York, 1955.

W. T. Reid, *Ordinary Differential Equations*, John Wiley and Sons, New York, 1971.

P. Hartman, *Ordinary Differential Equations*, John Wiley and Sons, New York, 1964.

Section 4

This material follows very closely that given in Ince. Very few of the modern books treat the Sturmian Theory in any detail. This is unfortunate, since the theory is so important to a thorough understanding of eigenvalue problems and the uniqueness theory of boundary-value problems.

Sections 5 and 6

The best discussion on boundary value problems is found in

P. B. Bailey, L. F. Shampine and P. E. Waltman, *Nonlinear Two Point Boundary Value Problems*, Academic Press, New York, 1968.

Section 6

There are a number of books which treat Green's functions in detail. For example, see

R. Courant and D. Hilbert, *Methods of Mathematical Physics* Vol. 1, Interscience, Inc., New York, 1953.

M. D. Greenberg, *Application of Green's Functions in Science and Engineering*, Prentice-Hall, Inc., Englewood Cliffs, N.J., 1971.

Section 7

The Riccati equation occurs in many areas of mathematical physics. Consequently, there are numerous books and papers concerning it. A few of these are

H. T. Davis, *Introduction to Non-Linear Differential and Integral Equations*, Dover, New York, 1962.

R. M. Redheffer, "On Solutions of Riccati's Equation as Functions of Initial Values," *J. Rat. Mech. Analysis*, **5** (1956), 835–848.

R. M. Redheffer, "The Riccati Equation: Initial Values and Inequalities," *Math. Ann.* **133** (1957), 235–250.

G. N. Watson, *Theory of Bessel Functions*, Cambridge Univ. Press, 1922; 2nd ed., 1944.

Section 8

See

W. T. Reid, *Riccati Differential Equations*, Academic Press, New York, 1972.

REVIEW OF NUMERICAL SOLUTION OF
ORDINARY DIFFERENTIAL EQUATIONS

1 Introduction

Most of the interesting equations which occur in practice require that their solutions be obtained by numerical means. In this chapter some of the classical techniques for numerically solving differential equations are discussed. Section 2 is devoted to the basic definitions of truncation error and roundoff errors. Also included is a summary of the advantages and disadvantages of one-step and multistep methods. The next three sections present, respectively, the classical Taylor's method, Runge-Kutta, and multistep methods. Sections 6 and 7 are devoted to the important concept of stability. Three types of instabilities are introduced: inherent, partial, and parasitic. Sections 8 and 9 discuss techniques for solving boundary value problems.

2 Definitions and Notation

The interval $[a, b]$ over which the independent variable x is defined will be divided into subintervals. The value of the true solution $y(x)$ is approximated at $n + 1$ evenly spaced points $(x_0, x_1, ..., x_n)$. Thus, the *mesh* or *step size* is given by

$$h = \frac{b - a}{n} \qquad \text{(II.2.1)}$$

and

$$x_i = x_0 + ih \qquad i = 1, ..., n \qquad \text{(II.2.2)}$$

Let the *true*, or *exact*, solution $y(x)$ at the mesh points be denoted by $y(x_i)$ and the *approximate* solution be denoted by y_i; so thus,

$$f_i = f(x_i, y) \sim f[x, y(x)]. \tag{II.2.3}$$

The difference between the approximate solution and the true solution, ignoring roundoff error, is called the *discretization* or *truncation error*, ε_i:

$$\varepsilon_i = y - y(x_i). \tag{II.2.4}$$

The *local truncation error* is the error obtained by integrating a differential equation across one step. This error is determined only by the particular method used to approximate the differential equation and does not depend upon the type of computer used.

The *roundoff error* is a function of the word size of the computer and is difficult to analyze; therefore, we shall ignore it.

The numerical procedures we shall study may be classified into two groups: the *one-step* and the *multistep* methods. A method is called a one-step method if y_{i+1} can be calculated with only the knowledge of y_i being necessary. Multistep methods require several values of y_i' and y_i (i.e., f_i).

One-Step Methods

A. Advantages
1. Are self-starting – only a knowledge of y_0 is needed to calculate y_1, etc.
2. Have easy-to-change step size
B. Disadvantages
1. Have large number of function evaluations
2. May need small step size for desired accuracy because of high truncation error; hence, is slow

Multistep Methods

A. Advantages
1. Have generally lower truncation error than one-step methods
2. Usually require fewer function evaluations than one-step methods
3. For comparable accuracy, usually are faster
B. Disadvantages
1. Are not self-starting. Usually a one-step method must be used to supply the missing values of y_j and y_j'
2. Do not have easy-to-change step size
3. May be more unstable

All of the methods we shall study for solving a first order differential equation are based upon: (1) a direct or indirect use of Taylor's expansion or (2) a use of closed- or open integration formulae of the Newton-Cotes type.

3   Taylor's Method

Consider the problem

$$y' = f[x, y(x)], \quad y(x_0) = \alpha. \tag{II.3.1}$$

We first use the Taylor expansion in the form

$$y(x + h) = y(x) + hy'(x) + \frac{h^2}{2!} y''(x) + \frac{h^3}{3!} y'''(x) + \cdots$$

$$+ \frac{h^k}{k!} y^{(k)}(x) + \frac{h^{k+1}}{(k+1)!} y^{(k+1)}(\xi), \tag{II.3.2}$$

where ξ lies between x and $x + h$. In order to use the above formula we must express the derivatives of y in terms of the total derivatives of f. That is,

$$y'(x) = f[x, y(x)]$$
$$y''(x) = f_x + f_y y' = f_x + f_y f = f'$$
$$y'''(x) = f_{xx} + f_{xy} f + f_y (f_x + f_y f) + f(f_{xy} + f_{yy} f)$$
$$= f_{xx} + f_{xy} f + f_y f_x + f_y^2 f + f f_{xy} + f^2 f_{yy}$$
$$= f_{xx} + 2f f_{xy} + f^2 f_{yy} + f f_y^2 + f_x f_y = f'',$$

and, in general, we have

$$y^{(n)} = \frac{d^{n-1} f[x, y(x)]}{dx^{n-1}}, \quad n = 1, 2, \ldots. \tag{II.3.3}$$

Substitution of these equations into (II.3.2) yields

$$y(x + h) = y(x) + hf[x, y(x)] + \frac{h^2}{2!} f'[x, y(x)] + \cdots$$

$$+ \frac{h^k}{k!} f^{(k-1)}[x, y(x)] + \frac{h^{k+1}}{(k+1)!} f^{(k)}[\xi, y(\xi)],$$

$$= y(x) + hT_k(x, y; h) + \frac{h^{k+1}}{(k+1)!} f^{(k)}[\xi, y(\xi)],$$

where

$$T_k(x, y; h) = f[x, y(x)] + \frac{h}{2!} f'[x, y(x)] + \cdots + \frac{h^{k-1}}{k!} f^{(k-1)}[x, y(x)].$$

The foregoing equations suggest the following numerical algorithm (Taylor's algorithm of order k)

$$y_{i+1} = y_i + hT_k(x_i, y_i; h),$$ (II.3.4)
$$y_0 = \alpha,$$

where

$$T_k(x_i, y_i; h) = f(x_i, y_i) + \frac{h}{2!} f'(x_i, y_i) + \ldots + \frac{h^{k-1}}{k!} f^{(k-1)}(x_i, y_i)$$

and

$$x_i = x_0 + ih, \quad i = 1, 2, \ldots$$

Taylor's method is conceptually very simple and a nice error bound is given; however, it is very difficult to use in practice because of the difficulty in taking the partial derivatives of f.

The case $k = 1$ is normally called Euler's method:

$$y_{i+1} = y_i + hf(x_i, y_i)$$
$$y_0 = \alpha.$$ (II.3.5)

4   Runge-Kutta Methods

In the previous section we discussed one of the most obvious techniques for solving ordinary differential equations. We found that Taylor's method was not of practical use, since the partial derivatives may be quite difficult to obtain. In this section we shall develop variations of Taylor's method which do not involve the evaluation of the partial derivatives of f and, hence, are the practical methods for the numerical solution of ordinary differential equations. These are called the *Runge-Kutta* methods. The Runge-Kutta procedures are designed to agree with the corresponding Taylor's expansion through terms in h^k, where k is the order of the method.

Let us now consider the derivation of the general second order Runge-Kutta methods. The derivation of the higher order formulas is analogous to the one which follows, but involves a great deal more arithmetic.

Let $k(x_i, y_i; h)$ be a weighted average of two derivative evaluations, k_1 and k_2, on the interval $x_i \leqslant x \leqslant x_{i+1}$; that is,

$$K(x_i, y_i; h) = ak_1 + bk_2, \qquad \text{(II.4.1)}$$

which gives the algorithm

$$y_{i+1} = y_i + h(ak_1 + bk_2). \qquad \text{(II.4.2)}$$

Let

$$\begin{aligned}
k_1 &= f(x_i, y_i) \\
k_2 &= f[x_i + ph, y_i + qhf(x_i, y_i)] \\
&= f(x_i + ph, y_i + qhk_1),
\end{aligned} \qquad \text{(II.4.3)}$$

where a, b, p, and q are to be established later.

The first few terms of the Taylor's series expansion for a function of two variables are

$$\begin{aligned}
f(x + r, y + s) = f(x, y) &+ rf_x(x, y) + sf_y(x, y) + r^2 f_{xx}(x, y)/2 \\
&+ s^2 f_{yy}(x, y)/2 + rs f_{xy}(x, y) \\
&+ 0[(|r| + |s|)^3].
\end{aligned} \qquad \text{(II.4.4)}$$

Now, using the foregoing formula, we expand k_2 in a Taylor's expansion and keep only those terms with powers of h less than or equal to 1:

$$\begin{aligned}
k_2 &= f[x_i + ph, y_i + qhf(x_i, y_i)] \\
&= f(x_i, y_i) + phf_x(x_i, y_i) + qhf(x_i, y_i) f_y(x_i, y_i) + 0(h^2).
\end{aligned} \qquad \text{(II.4.5)}$$

Using (II.4.3) and (II.4.5), we can rewrite (II.4.2) as

$$\begin{aligned}
y_{i+1} = y_i &+ h\{af(x_i, y_i) + bf(x_i, y_i)\} + h^2\{bpf_x(x_i, y_i) \\
&+ bqf(x_i, y_i) f_y(x_i, y_i)\} + 0(h^3).
\end{aligned}$$

Next, we expand the solution $y(x)$ in a Taylor series about x_i:

$$\begin{aligned}
y(x_i + h) = y(x_{i+1}) = y(x_i) &+ hf[x_i, y(x_i)] + \frac{h^2}{2} f'[x_i, y(x_i)] \\
&+ \frac{h^3}{3!} f''[\xi, y(\xi)], \qquad x < \xi < x_{i+1}.
\end{aligned}$$

Recall that

$$f'[x_i, y(x_i)] = f_x[x_i, y(x_i)] + f_y[x_i, y(x_i)] f[x_i, y(x_i)].$$

Equating the coefficients in like powers of h and assuming $y_i = y(x_i)$, we find that

$$a + b = 1, \qquad bp = \tfrac{1}{2}, \qquad bq = \tfrac{1}{2}.$$

Since we have three equations and four unknowns, the system is undetermined and one variable is to be chosen arbitrarily. Two of the most popular second-order methods (the improved Euler's method and the modified Euler's method) use $b = \tfrac{1}{2}$ and $b = 1$, respectively.

The most common fourth order formulas are:
(i) the classical fourth order Runge-Kutta

$$y_{i+1} = y_i + \tfrac{1}{6}h(k_1 + 2k_2 + 2k_3 + k_4), \qquad (\text{II.4.6})$$

where

$$k_1 = f(x_i, y_i)$$
$$k_2 = f(x_i + \tfrac{1}{2}h, y_i + \tfrac{1}{2}hk_1)$$
$$k_3 = f(x_i + \tfrac{1}{2}h, y_i + \tfrac{1}{2}hk_2)$$
$$k_4 = f(x_i + h, y_i + hk_3),$$

(ii) the fourth order Runge-Kutta scheme of Gill

$$y_{i+1} = y_i + \tfrac{1}{6}h\left(k_1 + 2\left(1 - \frac{1}{\sqrt{2}}\right)k_2 + 2\left(1 + \frac{1}{\sqrt{2}}\right)k_3 + k_4\right), \qquad (\text{II.4.7})$$

where

$$k_1 = f(x_i, y_i)$$
$$k_2 = f(x_i + \tfrac{1}{2}h, y_i + \tfrac{1}{2}h)$$
$$k_3 = f\left[x_i + \tfrac{1}{2}h, y_i + \left(-\tfrac{1}{2} + \frac{1}{\sqrt{2}}\right)hk_1 + \left(1 - \frac{1}{\sqrt{2}}\right)hk_2\right]$$
$$k_4 = f\left[x_i + h, y_i - \frac{1}{\sqrt{2}}hk_2 + \left(1 + \frac{1}{\sqrt{2}}\right)hk_3\right].$$

The fourth order formulas are probably the most widely used methods for solving differential equations, with Gill's method more popular than the classical method.

5   Multistep Methods

The Runge-Kutta methods are called one-step methods since, to calculate y_{i+1}, we need only a knowledge of y_i. This is desirable, for it means that the Runge-Kutta methods are self-starting and the step size can be easily adjusted during the computation. However, for the second order and higher methods, several intermediate evaluations of the function are necessary. These can become quite time consuming and, as a result, we look for other methods. One obvious way to proceed is to use previous values of y_i. However, by doing this, we lose two very desirable advantages: the self-starting feature and the ability to change the step size easily.

Again we consider the problem

$$
\begin{aligned}
y' &= f\left(x,\, y\left(x\right)\right), \\
y\left(0\right) &= \alpha.
\end{aligned}
\tag{II.5.1}
$$

We shall develop multistep algorithms based upon integration formulae of the type

$$
y_{i+1} = y_{i-p} + \int_{x_{i-p}}^{x_{i+1}} f\left(x,\, y\left(x\right)\right) dx.
\tag{II.5.2}
$$

Many different schemes may be developed by choosing various values of $p = 0, 1, \ldots, i$.

There are basically two types of quadrature formulae for equally spaced mesh points:

1. open on the right ($n+1$ nodes)

$$
x_i,\, x_{i-1},\, \ldots,\, x_{i-n}
$$

2. closed on the right ($n+1$ nodes)

$$
x_{i+1},\, x_i,\, x_{i-1},\, \ldots,\, x_{i+1-n}.
$$

Thus, the algorithms suggested by these two classes of methods are

$$
y_{i+1} = y_{i-p} + h \sum_{j=0}^{n} \alpha_j f\left(x_{i-j},\, y_{i-j}\right) \quad \text{(open)}
\tag{II.5.3}
$$

$$
y_{i+1} = y_{i-p} + h \sum_{j=0}^{n} \beta_j f\left(x_{i+1-j},\, y_{i+1-j}\right) \quad \text{(closed)}
\tag{II.5.4}
$$

By choosing a uniform mesh spacing, the coefficients α_j and β_j are independent of i. When the integers p and n are specified, the coefficients are then determined by the Newton-Cotes integration formulae.

Since a polynomial is easy to integrate, we approximate the function $f(x, y(x))$ over the $n+1$ nodes by an nth order degree polynomial. There are a number of equivalent ways to approach this problem, but perhaps the simplest is to use the Lagrange polynomials. Thus, we write

$$\int_{x_{i-p}}^{x_{i+1}} f(x, y(x))\, dx = \int_{x_{i-p}}^{x_{i+1}} P_n(x)\, dx + \text{error term (E.T.)}, \qquad \text{(II.5.5)}$$

where

$$P_n(x) = \sum_{j=0}^{n} l_j(x)\, f(x_{i+q-j}, y_{i+q-j}) \quad \begin{cases} q = 0 & \text{open} \\ q = 1 & \text{closed} \end{cases}$$

and

$$l_j(x) = \prod_{\substack{k=0 \\ k \neq j}}^{n} \frac{x - x_{i+q-k}}{x_{i+q-j} - x_{i+q-k}}.$$

Keep in mind that the node points in the formulae for $l_j(x)$ are different for the open and closed formulae. Then (II.5.5) can be written as

$$\int_{x_{i-p}}^{x_{i+1}} f(x, y(x))\, dx = \sum_{j=0}^{n} \left\{ \int_{x_{i-p}}^{x_{i+1}} l_j(x)\, dx \right\} f(x_{i+q-j}, y_{i+q-j}) + \text{E.T.} \quad \text{(II.5.6)}$$

Let

$$A_j = \int_{x_{i-p}}^{x_{i+1}} l_j(x)\, dx = \int_{x_{i-p}}^{x_{i+1}} \prod_{\substack{k=0 \\ k \neq j}}^{n} \frac{x - x_{i+q-k}}{x_{i+q-j} - x_{i+q-k}}\, dx.$$

If the nodal points $x_i, x_{i-1}, \ldots, x_{i-n}$ are used in the foregoing formula, $A_j = \alpha_j h$; if the nodal points $x_{i+1}, x_i, x_{i-1}, \ldots, x_{i+1-n}$ are used, $A_j = \beta_j h$. Let $x = x_{i-p} + sh$, $0 \leqslant s \leqslant 1+p$. Then

$$\alpha_j = \int_{-p}^{1} \prod_{\substack{k=0 \\ k \neq j}}^{n} \frac{s+k}{k-j}\, ds, \qquad \text{(II.5.7)}$$

and

$$\beta_j = \int_{-p}^{1} \prod_{\substack{k=0 \\ k \neq j}}^{n} \frac{s+k-1}{k-j}\, ds. \qquad \text{(II.5.8)}$$

The fundamental difference between the open and the closed formulae is the manner in which they must be solved. The open formula (II.5.3) is said to be an *explicit method* and is easily solved. However, since y_{i+1} appears on both sides of (II.5.4), the closed formula is said to be an *implicit method* and must be solved by iteration.

The normal procedure is to *predict* with an open formula and then to *correct* with a closed formula. That is, we use

$$y_{i+1}^{(0)} = y_{i-p} + h \sum_{j=0}^{n} \alpha_j f\left(x_{i-j}, y_{i-j}\right) \qquad \text{(II.5.9)}$$

to predict a value of y_{i+1} and then use

$$y_{i+1}^{(1)} = y_{i-p} + h \sum_{j=1}^{n} \beta_j f\left(x_{i+1-j}, y_{i+1-j}\right) + h\beta_0 f\left(x_{i+1}, y_{i+1}^{(0)}\right)$$

$$\text{(II.5.10)}$$

to correct the value. If desired, the corrector may be iterated a number of times to gain greater accuracy, but in practice only one or two iterations are ever used. Notice that the combination of (II.5.9) and (II.5.10) involves only two function evaluations per step. Each additional application of the corrector then involves one function evaluation.

In practice, the order of the predictor should be of the same order as the corrector or perhaps one less. This is fairly obvious, since we certainly would not want to correct with a procedure that is less accurate than the predictor; however, the more accurate the predictor, the faster we have convergence.

One of the most popular of the predictor-corrector schemes is that of the Adams-Moulton scheme

$$y_{i+1}^{(0)} = y_i + \frac{h}{24} \left\{ 55 f(x_i, y_i) - 59f\left(x_{i-1}, y_{i-1}\right) + 37f\left(x_{i-2}, y_{i-2}\right) \right.$$

$$\left. - 9f\left(x_{i-3}, y_{i-3}\right) \right\}. \qquad \text{(II.5.11)}$$

$$y_{i+1} = y_i + \frac{h}{24} \left\{ 9f\left(x_{i+1}, y_{i+1}^{(0)}\right) + 19f(x_i, y_i) - 5f\left(x_{i-1}, y_{i-1}\right) \right.$$

$$\left. + f\left(x_{i-2}, y_{i-2}\right) \right\}. \qquad \text{(II.5.12)}$$

The coefficients may be easily calculated from (II.5.7) and (II.5.8) by letting $p=0$ and $n=3$ (see exercise 2). The local truncation error for the Adams-Moulton method is $0(h^5)$. Notice that it requires only two function evaluation per step as compared to four function evaluations per step for the fourth order

Runge-Kutta scheme. Also notice that we must provide values for y_0, y_1, y_2, and y_3 before using the Adams-Moulton procedure. This is normally accomplished by starting with the fourth-order Runge-Kutta to provide values for y_0, y_1, y_2, and y_3 and then switching to the Adams-Moulton scheme.

6 Stability

The word *instability* is one of the most overworked words in the English language. It appears in the literature with many qualifying adjectives (e.g. inherent, partial, parasitic, strong, weak, etc.). Basically, a solution is said to be *unstable* if any errors introduced at some stage of the computation (e.g. from erroneous initial conditions, truncation error, or roundoff error) are propagated without bound throughout subsequent calculations.

We shall consider three types of instability: (1) *Inherent instability*, which is introduced by the differential equations and the initial or boundary conditions, has nothing to do with the particular numerical scheme being used. (See exercises 6–11.) This is the type of instability that we try to overcome by using invariant imbedding, discussed in later chapters, (2) *Parasitic instability* is introduced by increasing the order of differencing in the use of multistep methods. (See exercises 3–5.) (3) *Partial instability* is a function of the particular differential equation being solved, its initial or boundary conditions, and the numerical algorithm (usually a one-step procedure) being used.

Exercises 3–11, although certainly not complete, give the reader some idea of how the first two types of instability can be encountered. Inherent instability will be discussed in more detail in section 7 and in later chapters, but parasitic instability more properly belongs to a course in numerical analysis.

The phenomenon of partial instability can be seen easily from examining Euler's method. The total error at x_{i+1} is related to the total error at x_i by

$$\varepsilon_{i+1} = \varepsilon_i + h\{f(x_i, y_i) - f[x_i, y(x_i)]\} - \frac{h^2}{2} f'[\xi, y(\xi)], \quad \text{(II.6.1)}$$

where $x_i < \xi < x_{i+1}$. From the mean value theorem, we have

$$f(x_i, y_i) - f[x_i, y(x_i)] = [y_i - y(x_i)] \frac{\partial f}{\partial y}\bigg|_{x_i, \alpha}, \quad \text{(II.6.2)}$$

where α is between y_i and $y(x_i)$. Thus,

$$\varepsilon_{i+1} = \varepsilon_i \left(1 + h \frac{\partial f}{\partial y}\bigg|_{x_i, \alpha}\right) - \frac{h^2}{2} f'[\xi, y(\xi)] \quad \text{(II.6.3)}$$

The coefficient of ε_i is the important term in determining the stability. If $[1 + h(\partial f/\partial y)] < 1$, then the error will tend to die out. However, if $[1 + h(\partial f/\partial y)] > 1$ (i.e. $\partial f/\partial y > 0$), then the error will be amplified as i increases. Thus, the partial instability is a function of the particular equation, the particular algorithm, and the step size and can be controlled to a certain extent by appropriate choice of the step size.

It is necessary to realize that, if the solution is increasing, the *absolute* value of the error is, in general, not so important as the *relative* error. For example, consider the equation

$$y' = y, \qquad y(0) = \alpha, \tag{II.6.4}$$

which has the solution

$$y(z) = \alpha e^z, \tag{II.6.5}$$

and the equation (with a slightly perturbed initial condition)

$$w' = w, \qquad w(0) = \alpha + \varepsilon, \tag{II.6.6}$$

which has the solution

$$w(z) = \alpha e^z + \varepsilon e^z. \tag{II.6.7}$$

The absolute difference between $y(z)$ and $w(z)$ is

$$|y(z) - w(z)| = |\varepsilon|\, e^z,$$

which is unbounded. However, since the solution is also unbounded, the relative error is given by

$$\left| \frac{y(z) - w(z)}{y(z)} \right| = \left| \frac{\varepsilon}{\alpha} \right|,$$

and other errors (such as truncation and roundoff) will not grow appreciably. Hence, this equation can be solved accurately, in a relative sense, even for large z.

There are a number of other techniques which are satisfactory for solving initial-value problems, such as Hamming's method, which we simply do not have time or space to elaborate upon. Notice that the methods discussed thus far do not require that the differential equation be linear. This is typical of the methods for solving initial-value problems, but, as we shall see in the next section, it is not true of all methods for solving boundary-value problems.

7   Inherent Instability and Backward Integration

In order to illustrate the phenomenon of inherent instability, we shall examine the difference between calculating the dominant solution and the minimal solution of a second-order homogeneous equation. Let us consider the two problems

$$y_1''(z) - y_1(z) = 0, \quad y_1(0) = 1 + \varepsilon, \quad y'(0) = 1, \quad \text{(II.7.1)}$$

and

$$y_2''(z) - y_2(z) = 0, \quad y_2(0) = 1 + \varepsilon, \quad y'(0) = -1, \quad \text{(II.7.2)}$$

where ε is a small positive number used to simulate an error or perturbation. The solutions of (II.7.1) and (II.7.2) are

$$y_1(z) = (1 + \tfrac{1}{2}\varepsilon) e^z + \tfrac{1}{2}\varepsilon e^{-z} \quad \text{(II.7.3)}$$

and

$$y_2(z) = \tfrac{1}{2}\varepsilon e^z + (1 + \tfrac{1}{2}\varepsilon) e^{-z}. \quad \text{(II.7.4)}$$

The solutions that we really desire are, of course, the unperturbed solutions, namely, when $\varepsilon = 0$.

The relative errors between the perturbed and unperturbed solutions are given by

$$r_1 = \left| \frac{y_1(z, \varepsilon) - y_1(z, 0)}{y_1(z, 0)} \right| = \frac{\tfrac{1}{2}\varepsilon e^z + \tfrac{1}{2}\varepsilon e^{-z}}{e^z} = \tfrac{1}{2}\varepsilon (1 + e^{-2z}) \quad \text{(II.7.5)}$$

and

$$r_2 = \left| \frac{y_2(z, \varepsilon) - y_2(z, 0)}{y_2(z, 0)} \right| = \frac{\tfrac{1}{2}\varepsilon e^z + \tfrac{1}{2}\varepsilon e^{-z}}{e^{-z}} = \tfrac{1}{2}\varepsilon (e^{2z} + 1). \quad \text{(II.7.6)}$$

In the first case, when we are trying to obtain the dominant solution, e^z, the relative error is a bounded function; in the second case, when we are trying to obtain the minimal solution, e^{-z}, the relative error grows exponentially.

Perhaps this can best be understood by graphing some of the pertinent functions involved. Let

$$u_1(z) = (1 + \tfrac{1}{2}\varepsilon) e^z, \quad \text{(II.7.7)}$$

$$v_1(z) = \tfrac{1}{2}\varepsilon e^{-z}. \quad \text{(II.7.8)}$$

Then

$$y_1(z) = u_1(z) + v_1(z). \quad \text{(II.7.9)}$$

In Figure 2-1 we have graphed $u_1(z)$, $v_1(z)$, $u_1(z) + v_1(z)$ and the desired solution e^z. Notice that the computed solution $y_1(z)$ approximates e^z quite nicely,

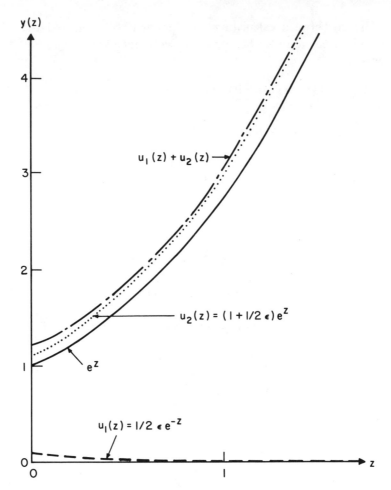

Figure 2-1. Comparison of Perturbed and Unperturbed Solutions of
$y'' - y = 0,\ y(0) = 1 + \varepsilon,\ y'(0) = 1$

as predicted by the fact that the relative error is bounded by ε. In fact, the relative error decreases as z grows.

In Figure 2-2 we have graphed $u_2(z)$, $v_2(z)$, $u_2(z) + v_2(z)$ and the desired solution e^{-z}, where $u_2(z) = \frac{1}{2}\varepsilon e^z$ and $v_2(z) = (1 + \frac{1}{2}\varepsilon)\, e^{-z}$. In this case the computed solution $y_2(z)$ follows the desired solution for only a short interval and then deviates drastically. This is, of course, as predicted since the relative error grows exponentially.

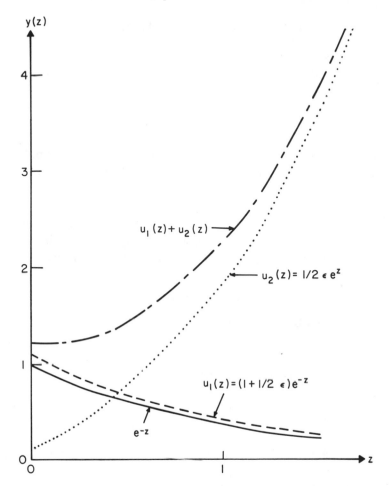

Figure 2-2. Comparison of Perturbed and Unperturbed Solutions of $y'' - y = 0$, $y(0) = 1 + \varepsilon$, $y'(0) = -1$

We now turn to a technique for overcoming the instability encountered in trying to calculate the minimal solution, e^{-z}. Our foregoing problem stemmed from the fact that the general solution of our second order equation had complementary solutions e^{z} and e^{-z}. We found that we could compute the growing solution quite accurately. However, the decaying solution could not be computed very accurately since any error, regardless of how small, introduced a component of the growing solution which eventually swamped e^{-z}. Thus, if we could in some way integrate in the backward direction, then e^{-z} would become the growing solution and e^z would become the decaying solution. We shall proceed to discuss such a method.

We consider the equation

$$\hat{u}''(z) - \hat{u}(z) = 0 \tag{II.7.10}$$

subject to

$$\hat{u}(X) = 1, \quad \hat{u}'(X) = 0, \quad X \gg 0. \tag{II.7.11}$$

(We could have chosen other initial conditions, but this choice simplifies some of the arithmetic involved.) The solution of (II.7.10) subject to (II.7.11) is

$$\hat{u}(z) = \tfrac{1}{2}e^{-X}e^z + \tfrac{1}{2}e^X e^{-z}. \tag{II.7.12}$$

Keep in mind that we wish to approximate e^{-z} over some desired interval. Since we chose the initial conditions at X arbitrarily, we must normalize all of our values to agree with those of $y(z)$, the desired solution. This can be accomplished by defining

$$u(z) = y(0)\,\hat{u}(z)/\hat{u}(0). \tag{II.7.13}$$

Since $y(0) = 1$ and $\hat{u}(0) \approx \tfrac{1}{2}e^X$, we have

$$u(z) \cong e^{-2X}e^z + e^{-z} \tag{II.7.14}$$

The relative error between $y(z)$ and $u(z)$ is

$$r = \left| \frac{e^{-2X}e^z + e^{-z} - e^{-z}}{e^{-z}} \right| = e^{-2(X-z)} \tag{II.7.15}$$

In this simple case we can actually compute the interval I over which $u(z)$ is a good approximation of $y(z)$. Suppose we wish to have the relative error to be less than ε over I. To find I we must use (II.7.15) and we find

$$e^{-2(X-z)} < \varepsilon \tag{II.7.16}$$

or

$$z < X + \tfrac{1}{2}\ln \varepsilon. \tag{II.7.17}$$

Thus, if we choose $I = X + \tfrac{1}{2}\ln\varepsilon$, the relative error will be less than ε over I. For $X = 10$ and $\varepsilon = 0.01$, then $I = 6.55$.

If I is too short, then to increase the interval we need only choose a larger X and repeat the process until the values of $u(z)$ converge over I. (This is the process which must be used on a computer since, in general, we have no *a priori* way of estimating the interval I.) Again, a graph should illustrate the ideas involved. Pick $X = X_1$, let $u_1(z) = \tfrac{1}{2}e^{-X}e^z$ and $u_2(z) = \tfrac{1}{2}e^X e^{-z}$ and graph $u_1(z)$, $u_2(z)$ and $u_1(z) + u_2(z)$. (See Figure 2-3.) Observe that $u_2(z)$ and $u_1(z) + u_2(z)$ agree quite

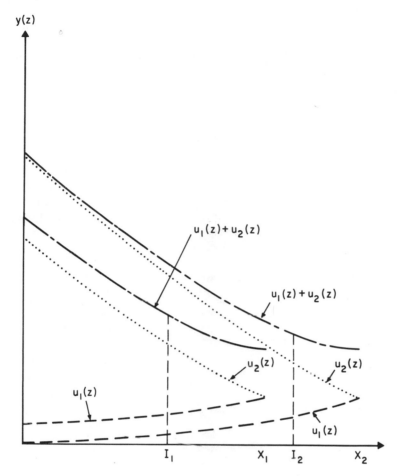

Figure 2-3.   Backward Integration for Various Values of X

closely for $0 \leqslant z \leqslant I_1$. In order to extend this to an interval I_2, we pick another $X_2 > X_1$ and repeat the process and then normalize all values.

Notice that in the forward direction we have no way of extending the interval of good approximation since ε is fixed; whereas, in the backward direction we need only pick larger and larger values for X.

As an actual example let us consider the application of backward integration to the equation

$$y''(z) - (z^2 - 1)\, y(z) = 0, \tag{II.7.18}$$

$$y(0) = 1, \quad y'(0) = 0. \tag{II.7.19}$$

The general solution of (II.7.18) is

$$y(z) = Ae^{-z^2/2} + Be^{-z^2/2} \int_0^z e^{t^2} \, dt \qquad (II.7.20)$$

and with the initial conditions (II.7.19), the desired solution is

$$y(z) = e^{-z^2/2}, \qquad (II.7.21)$$

which is clearly the minimal solution. Thus, standard initial-value techniques will have difficulty in accurately approximating the solution over an interval of appreciable length. Also, since $e^{-z^2/2}$ is dominant with decreasing z, the backward integration scheme should be feasible. The results of such an experiment are given in Table II.2-1.

TABLE II.2-1.   *Results of Forward and Backward Integration for $y'' - (z^2-1)y=0$, $y(0)=1$, $y'(0)=0$*

z	$y(z)$ Exact	Forward Integration	Backward Integration
0	1.0	1.0	1.0
1	6.065307 − 01	6.065307 − 01	6.065306 − 01
2	1.353353 − 01	1.353353 − 01	1.353353 − 01
3	1.110900 − 02	1.110900 − 02	1.110900 − 02
4	3.354626 − 04	3.354648 − 04	3.354627 − 04
5	3.726653 − 06	3.879808 − 06	3.726653 − 06
6	1.522998 − 08	3.103631 − 05	1.522998 − 08
7	2.289735 − 11	1.761642 − 02	2.289736 − 11
8	1.266417 − 14	2.780004 − 01	1.266417 − 14
9	2.576757 − 18	1.212432 + 05	2.576759 − 18
10	1.928750 − 22	1.456040 + 09	1.928756 − 22

8 Linear Boundary-Value Problems – Initial-Value Techniques

Boundary-value problems are, in general, much more difficult to analyze than initial-value problems. First of all, the questions of existence and uniqueness are more difficult to answer. We shall assume, unless otherwise stated, that the solution of each boundary-value problem under consideration is unique. Also, in contrast to the initial-value problems, there are only a few techniques which apply equally well to linear and nonlinear problems.

The general linear boundary-value problem may be written as

$$y''(z) + f_1(z) y'(z) + f_2(z) y(z) + f_3(z) = 0,$$
$$\alpha_1 y(a) + \beta_1 y'(a) + \gamma_1 y(b) + \delta_1 y'(b) = \eta_1, \tag{II.8.1}$$
$$\alpha_2 y(a) + \beta_2 y'(a) + \gamma_2 y(b) + \delta_2 y'(b) = \eta_2.$$

Although the techniques we shall discuss can be easily modified to handle the general linear problem, we shall, for sake of convenience, develop schemes for the problem

$$y''(z) = f(z) y'(z) + g(z) y(z) + h(z) \tag{II.8.2}$$

$$y(0) = \alpha_1, \quad y(1) = \alpha_2 \tag{II.8.3}$$

Recall that the solution of (II.8.2) can be expressed as

$$y(z) = A y_1(z) + B y_2(z) + y_p(z), \tag{II.8.4}$$

where $y_1(z)$ and $y_2(z)$ are any two linearly independent solutions of the homogeneous portion of (II.8.2) and $y_p(z)$ is a particular solution of the nonhomogeneous equation. The constants A and B are to be adjusted so that the solution of (II.8.2) is obtained. For example, if $y_1(x)$ is the solution of

$$y_1''(z) = f(z) y_1'(z) + g(z) y_1(z)$$
$$y_1(0) = 1, \quad y_1'(0) = 0 \tag{II.8.5}$$

$y_2(z)$ is the solution of

$$y_2''(z) = f(z) y_2'(z) + g(z) y_2(z)$$
$$y_2(0) = 0, \quad y_2'(0) = 1 ; \tag{II.8.6}$$

and $y_p(z)$ is the solution of

$$y_p''(z) = f(z) y_p'(z) + g(z) y_p(z) + h(z)$$
$$y_p(0) = 0, \quad y_p'(0) = 0, \tag{II.8.7}$$

we have

$$y(z) = A y_1(z) + B y_2(z) + y_p(z). \tag{II.8.8}$$

We obtain A and B from the boundary conditions of (II.8.3) and the initial conditions of (II.8.5) and (II.8.7). That is, if we evaluate (II.8.8) at $z=0$, we obtain

$$\alpha_1 = y(0) = A y_1(0) + B y_2(0) + y_p(0) = A. \tag{II.8.9}$$

At $z=1$, we find

$$\alpha_2 = y(1) = \alpha_1 y_1(1) + B y_2(1) + y_p(1)$$

or

$$B = \frac{\alpha_2 - \alpha_1 y_1(1) - y_p(1)}{y_2(1)}. \tag{II.8.10}$$

We must assume $y_2(1) \neq 0$. If it is zero, then the homogeneous problem has a nontrivial solution and, in general, the inhomogeneous problem has no solution. We will discuss this matter further when we get to eigenvalue problems. The initial conditions in (II.8.5) and (II.8.7) are somewhat arbitrary. We demand only that $y_1(z)$ and $y_2(z)$ be independent.

The number of integrations in the above procedure can be reduced from three to two by using the following formulation. Since the solution of a second-order linear differential equation is a linear combination of two linearly independent solutions, we write the solution of (II.8.2, II.8.3) as

$$y(z) = Ay_1(z) + By_2(z), \tag{II.8.11}$$

where $y_1(z)$ and $y_2(z)$ are solutions of

$$\begin{aligned} y_1''(z) &= f(z) y_1'(z) + g(z) y_1(z) + h(z) \\ y_1(0) &= \alpha_1, \quad y_1'(0) = \beta_1, \end{aligned} \tag{II 8.12}$$

and

$$\begin{aligned} y_2''(z) &= f(z) y_2'(z) + g(z) y_2(z) + h(z) \\ y_2(0) &= \alpha_1, \quad y_2'(0) = \beta_2. \end{aligned} \tag{II.8.13}$$

The constants A and B are then found from

$$\alpha_1 = y(0) = Ay_1(0) + By_2(0) = (A + B)\alpha_1 \tag{II.8.14}$$

and

$$\alpha_2 = y(1) = Ay_1(1) + By_2(1). \tag{II.8.15}$$

The first relation (II.8.14) yields

$$A + B = 1. \tag{II.8.16}$$

Both of the above techniques are incapable of dealing with inherent instabilities. Thus, we turn to invariant imbedding and boundary-value techniques.

9   Linear Boundary-Value Problems – Boundary-Value Techniques

Before developing any techniques, let us consider a few heuristic arguments for using boundary-value methods. Consider the problem

$$y''(z) = 100y(z),$$
$$y(0) = \alpha_1, \quad y(1) = \alpha_2. \tag{II.9.1}$$

This problem has the general solution

$$y(z) = Ae^{10z} + Be^{-10z}. \tag{II.9.2}$$

Thus,

$$\alpha_1 = y(0) = A + B,$$
$$\alpha_2 = y(1) = Ae^{10} + Be^{-10}. \tag{II.9.3}$$

From these two equations we find

$$A = \frac{\alpha_1 e^{-10} - \alpha_2}{e^{-10} - e^{10}}. \tag{II.9.4}$$

The sensitivity of A to changes in the boundary conditions is given by

$$\frac{dA}{d\alpha_1} = \frac{e^{-10}}{e^{-10} - e^{10}}, \quad \frac{dA}{d\alpha_2} = \frac{-1}{e^{-10} - e^{10}}. \tag{II.9.5}$$

Thus, A is not very sensitive to changes in the boundary conditions and, hence, we expect boundary-value techniques to be fairly good.

Now consider the initial-value problem

$$y''(z) = 100y(z),$$
$$y(0) = \alpha_1, \quad y'(0) = \beta_1. \tag{II.9.6}$$

Again, the general solution is given by

$$y(z) = Ce^{10z} + De^{-10z}. \tag{II.9.7}$$

From the initial conditions we find

$$C = \frac{10\alpha_1 + \beta_1}{20}. \tag{II.9.8}$$

The sensitivity of C to changes in the initial conditions is given by

$$\frac{dc}{d\alpha_1} = \frac{1}{2}, \quad \frac{dc}{d\beta_1} = \frac{1}{20}. \tag{II.9.9}$$

Hence, the coefficient of e^{10z} is much more sensitive to changes in the initial conditions than it is to changes in the boundary conditions.

Further preliminary discussion to the boundary-value techniques involves some finite difference approximations to derivatives. Consider a Taylor's series expansion:

$$y(z + h) = y(z) + hy'(z) + \frac{h^2}{2} y''(z) + \cdots. \qquad (\text{II.9.10})$$

Solving for $y'(z)$, we get

$$y'(z) = \frac{y(z + h) - y(z)}{h} - \frac{h}{2} y''(\xi) \qquad (\text{II.9.11})$$

where ξ lies between z and $z+h$. If we are willing to neglect the term $\frac{1}{2}hy''(\xi)$, we have as an approximation for $y'(z)$

$$y'(z) \approx \frac{y(z + h) - y(z)}{h}. \qquad (\text{II.9.12})$$

This is sometimes referred to as the *forward difference quotient*. (See Figure 2-4.)

Now consider the expansion

$$y(z - h) = y(z) - hy'(z) + \frac{h^2}{2} y''(z) - \cdots. \qquad (\text{II.9.13})$$

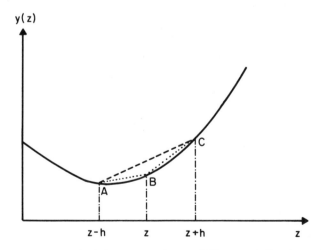

Figure 2-4.   Geometrical Interpretations of Forward, Backward and Central Difference Approximations of Derivative

Again, solving for $y'(z)$

$$y'(z) = \frac{y(z) - y(z-h)}{h} + \frac{h}{2} y''(\xi), \tag{II.9.14}$$

where ξ lies between z and $z-h$. Neglecting the term $\frac{1}{2}hy''(\xi)$, we obtain the *backward difference quotient*:

$$y'(z) \approx \frac{y(z) - y(z-h)}{h}. \tag{II.9.15}$$

If we subtract (II.9.13) from (II.9.10), we get

$$y(z+h) - y(z-h) = 2hy'(z) + \frac{1}{3}h^3 y^{(3)}(\xi), \tag{II.9.16}$$

where ξ lies between $x-h$ and $x+h$. Solving for $y'(x)$, we obtain

$$y'(z) = \frac{y(z+h) - y(z-h)}{2h} + \frac{1}{6}h^2 y^{(3)}(\xi). \tag{II.9.17}$$

Again, neglecting the small term $\frac{1}{6}h^2 y^{(3)}(\xi)$, which is now of order h^2, we get the *central difference quotient*:

$$y'(z) \approx \frac{y(z+h) - y(z-h)}{2h}. \tag{II.9.18}$$

Geometrically, the forward difference scheme says to take the slope of the chord BC as the derivative of $y(z)$ at z, the backward difference scheme says to use the slope of the chord AB, and the central difference scheme says to use the slope of the chord AC. (See Figure 2-4.)

We can also derive an approximation for $y''(z)$ in a similar manner. Add equations (II.9.10) and (II.9.13). We obtain

$$y(z+h) + y(z-h) = 2y(z) + h^2 y''(z) + \frac{1}{12} h^4 y^{(iv)}(\xi), \tag{II.9.19}$$

where ξ lies between $z-h$ and $z+h$. Solving for $y''(z)$, we get

$$y''(z) = \frac{y(z+h) - 2y(z) + y(z-h)}{h^2} - \frac{h^2}{12} y^{(iv)}(\xi). \tag{II.9.20}$$

Neglecting the error term $h^2/12\, y^{(iv)}(\xi)$, we obtain the *central difference scheme* for $y''(z)$:

$$y''(z) \approx \frac{y(z+h) - 2y(z) + y(z-h)}{h^2} \tag{II.9.21}$$

We are now in a position to develop a numerical scheme for the solution of

$$y''(z) - f(z)\, y'(z) - g(z)\, y(z) = k(z)$$
$$y(a) = \alpha_1, \quad y(b) = \alpha_2, \quad a \leqslant z \leqslant b. \tag{II.9.22}$$

The interval $a \leqslant z \leqslant b$ will be divided into N parts so that $h = (b-a)/(N+1)$. We now approximate the derivatives in (II.9.22) by the central difference quotients and obtain

$$\frac{y_{i-1} - 2y_i + y_{i+1}}{h^2} - f(z_i) \frac{y_{i+1} - y_{i-1}}{2h}$$
$$- g(z_i)\, y_i = k(z_i), \quad i = 1, 2, \ldots, N \tag{II.9.23}$$

with

$$y(a) = y_0 = \alpha_1, \quad y(b) = y_{N+1} = \alpha_2$$

Multiplying (II.9.23) by $-h^2/2$, we obtain

$$- b_i y_{i-1} + a_i y_i - c_i y_{i+1} = - h^2/2 k_i, \quad i = 1, 2, \ldots, N, \tag{II.9.24}$$

where

$$a_i = 1 + \tfrac{1}{2} h^2 g(z_i), \quad b_i = \tfrac{1}{2}\{1 + \tfrac{1}{2} h f(z_i)\}$$
$$c_i = \tfrac{1}{2}\{1 - \tfrac{1}{2} h f(z_i)\}.$$

We can now write (II.9.24) in matrix form as

$$Ay = d, \tag{II.9.25}$$

where

$$A = \begin{pmatrix}
a_1 & -c_1 & 0 \cdots & & & & 0 \\
-b_2 & a_2 & -c_2 & 0 \cdots & & & \vdots \\
0 & -b_3 & a_2 & -c_3 & & & \\
\vdots & & & & \ddots & & \\
0 & & & & -b_{N-1} & a_{N-1} & -c_{N-1} \\
& & & & & -b_N & a_N
\end{pmatrix}$$

$$y = \begin{pmatrix} y_1 \\ y_2 \\ \vdots \\ y_N \end{pmatrix}, \quad d = -\frac{h^2}{2}\begin{pmatrix} d_1 \\ d_2 \\ \vdots \\ d_N \end{pmatrix} + \begin{pmatrix} b_1\alpha_1 \\ 0 \\ \vdots \\ 0 \\ c_N\alpha_2 \end{pmatrix}.$$

Notice that the matrix A is tridiagonal. We have reduced the solution of the differential equation to the solution of a tridiagonal system of equations. There is a simple algorithm which is a variation of Gaussian elimination to solve the system.

Let

$$g_1 = c_1/b_1 \quad \text{and} \quad h_1 = d_1/b_1.$$

Then for

$$j = 2, 3, ..., N - 1$$
$$F = b_j - a_j g_{j-1},$$
$$h_j = (d_j - a_j h_{j-1})/F,$$
$$g_j = c_j/F.$$

The solution is then obtained by backward substitution

$$y_N = (d_N - a_N h_{N-1})/(b_N - a_N g_{N-1}),$$
$$y_j = h_j - g_j y_{j+1}, \quad j = N - 1, N - 2, ..., 1.$$

The above algorithm is very easy to program and requires only a little over $6N$ storage locations. The algorithm has several decided advantages over the standard Gaussian elimination. For example, the Gauss method takes approximately $\frac{1}{3}N(N^2+3N-1)$ multiplications, additions, and divisions, while this version has only $3N$ multiplications, $3N$ additions and $2N$ divisions. This not only is an advantage in speed, but also contributes to controlling the roundoff error. It can generalize to handle other specialized matrix systems.

The above procedure must be altered slightly if the boundary conditions involve the derivative. Assume that the boundary conditions for (II.9.22) are given by

$$y'(a) = \beta_1, \quad y'(b) = \beta_2. \tag{II.9.26}$$

Using the central difference quotient to approximate the derivative at $x=a$ and $x=b$, we have

$$\frac{y_1 - y_{-1}}{2h} = \beta_1, \quad \frac{y_{N+2} - y_N}{2h} = \beta_2. \tag{II.9.27}$$

The difference equations must be extended to include $i=0$ and $i=N+1$. This new set of equations will involve y_{-1} and y_{N+2}. We use (II.9.27) to eliminate these two values from the equations. Then the first and the last row of the matrix A become

$$a_0 y_0 - (b_0 + c_0) y_1 = -\frac{h^2}{2} k(x_0) - 2hb_0\beta_1$$

$$- (b_{N+1} + c_{N+1}) y_N + a_{N+1} y_{N+1} = 2h\beta_2 c_{N+1} - \frac{h^2}{2} k(x_{N+1}).$$

$$\text{(II.9.28)}$$

10 Exercises

1. Solve the ordinary differential equation

$$y''(z) = -4y(z),$$

   subject to the initial conditions $y(0)=1$, $y'(0)=0$, using the Taylor's expansion approach described in section 3, and compare with the analytical solution.

2. Use (II.5.7) and (II.5.8) with $p=0$ and $n=3$ to calculate the coefficients for the Adams-Moulton predictor-corrector scheme.

3. Consider the first-order equation $y'=ay$, $y(0)=1$. Discuss the possibility of solving this equation by differentiating and solving the resulting second-order problem. In terms of a numerical solution is there a significant difference between the case $a=+1$ and the case $a=-1$?

4. An obvious numerical technique for solving $y'=-y$, $y(0)=1$ is to approximate the derivative y' with the forward difference scheme $y' \approx (y(t+\Delta)-y(t))/\Delta$, which has a truncation error of order Δ. If we let $t=k\Delta$, then the resulting difference equation is $y_{k+1}+(\Delta-1)y_k=0$, $y_0=1$. Show that the solution of this difference equation is $y_k=(1-t/k)^k \approx e^{-t}$.

5. Consider the second-order approximation $y' \approx (y(t+\Delta)-y(t-\Delta))/2\Delta$. The resulting difference scheme is $y_{k+1}+2\Delta y_k - y_{k-1}=0$, with $y_0=1$. Use $y_1=1-\Delta+\Delta^2/2$. The difference equation has two complementary solutions of the form r_1^k and r_2^k, where r_1 and r_2 are the roots of $r^2+2\Delta r-1=0$. Show that $u_k \approx (1+\Delta^2/4) e^{-t}-(-1)^k \Delta^2/4 \, e^t$. Observe that, even though

our approximation of the derivative was more accurate, we introduced an extraneous solution as in exercise 3.

6. The general solution of the second-order equation $y''=y$ is $y(z)=Ae^z + Be^{-z}$. If we choose the initial conditions to be $y(0)=1$, $y'(0)=-1$, the solution is $y(z)=e^{-z}$. Discuss the solution if we perturb the initial condition $y(0)=1$ by ε, where ε is small, but could be positive or negative. That is, the initial conditions are now $y(0)=1+\varepsilon$ and $y'(0)=-1$. Compare the solutions at $z=10$ with $\varepsilon=10^{-3}$.

7. Since we know that the solution of the unperturbed problem in the above exercise tends to zero as $z\to\infty$, we might possibly be able to solve the problem by finding a $z=X\geqslant0$ and setting the initial conditions to be $y(X)=0$, $y'(X)=1$ and solve the problem backwards. Let $\hat{u}(z)$ denote the solution with the above initial conditions and define $u(z)=\hat{u}(z)/\hat{u}(0)$. (If $y(0)\neq1$, we need to set $u(z)=y(0)\,\hat{u}(z)/\hat{u}(0)$). Show that, given an $\varepsilon>0$, the relative error between $y(z)$ and $u(z)$ is less than ε on the interval $0\leqslant z\leqslant I$, where $I\approx X+\frac{1}{2}\ln\varepsilon$. For $X=10$, find the intervals for which the relative error is less than 0.01 and 0.001.

8. In the above exercise do we really need to use the initial conditions $y(X)=0$, $y'(X)=+1$, or could we simply use $y(X)=\alpha$, $y'(X)=\beta$, where $\alpha\neq\beta$? Show that $\alpha>\beta$ implies $u>y$ and $\alpha<\beta$ implies $u<y$ and then derive a new expression for I.

9. In exercise 6 we saw that perturbing the initial conditions slightly produced a drastic change in the solution. However, in exercise 8 we saw that we could use almost any initial conditions. Please clarify this apparent confusion.

10. Solve $y''-3y'+2y=0$, $y(0)=1$, $y'(0)=1$. Discuss the use of backward integration to solve this problem.

11. If the differential equation has variable coefficients, it may be hard to obtain a general solution and, hence, difficult to estimate X necessary for a reasonable interval, I. Discuss a procedure for backward integration for a general homogeneous, linear second-order equation.

12. Derive the algorithm presented in section 9 for solving the tridiagonal system of equations (II.9.24). (Hint: The algorithm is a simple variation of the Gaussian elimination procedure, which takes into account the regular spacing of the zeros of the matrix of coefficients.)

11 Bibliographical Discussion

Sections 4 and 5

Although many authors state that one of the advantages of the one-step methods is that the step size can be easily changed, very few state that deciding when to change can be somewhat difficult. A discussion of changing step size is beyond the scope of this book; however, there have been several papers which have discussed methods for changing step size based upon estimation of local truncation errors. See

F. Ceschino and J. Kuntzmann, *Numerical Solution of Initial Value Problems*, Prentice-Hall, Englewood Cliffs, New Jersey, 1966.

R. England, "Error Estimates for Runge-Kutta Type Solutions to Systems of Ordinary Differential Equations," *Computer Journal* **12** (1969), 166–170.

R. E. Huddleston, "Variable-Step Truncation Error Estimates for Runge-Kutta Methods of Order 4 or Less," *J. Math. Anal. Appl.* **39** (1972), 636–646.

R. E. Huddleston, "Selection of Step Size in the Variable Step Predictor-Corrector Method of Van Wyk," *J. Comp. Phys.* **9** (1972), 528–537.

The best references for numerical solution for ordinary differential equations are

P. Henrici, *Discrete Variable Methods in Ordinary Differential Equations*, John Wiley and Sons New York, 1962.

C. W. Gear, *Numerical Initial Value Problems in Ordinary Differential Equations*, Prentice-Hall. Inc., Englewood Cliffs, New Jersey.

Section 6

The stability of differential equations is discussed in

R. E. Bellman, *Methods of Nonlinear Analysis*, Academic Press, New York, 1970.

Excellent discussions of numerical stability may be found in

G. Dahlquist, "Convergence and Stability in the Numerical Solution of Ordinary Differential Equations," *Math. Scand.* **4** (1956), 33–53.

I. Babuska, M. Prager and E. Vitasek, *Numerical Processes in Differential Equations*, Interscience, New York, 1966.

L. Fox, *Numerical Solution of Ordinary and Partial Differential Equations*, Addison-Wesley Publ. Co., Inc., Reading, Mass., 1962.

Section 7

Although the idea of backward integration has been used quite extensively in solving difference equations it is rarely discussed in books on numerical solution of ordinary differential equations. See

D. E. Amos and J. W. Burgmeier, "Computation with Three-Term, Linear, Non-Homogeneous Recursion Relations," *SIAM Review* **15** (1973).

W. Gautschi, "Computational Aspects of Three-Term Recurrence Relations," *SIAM Review* **9** (1967), 24–82.

J. Oliver, "The Numerical Solution of Linear Recurrence Relations," *Numerische Mathematik* **11** (1968), 349–360.

M. R. Scott, "Numerical Solution of Unstable Initial-Value Problems by Invariant Imbedding," *The Computer Journal* **13** (1970), 397–400.

Section 8–9

For an excellent discussion of both initial-value (shooting techniques) and boundary-value techniques see

H. B. Keller, *Numerical Methods for Two-Point Boundary-Value Problems*, Blaisdell Publishing Co., Waltham, Mass., 1968.

Shooting techniques are discussed in detail in

S. M. Roberts and J. S. Shipman, *Two Point Boundary Value Problems: Shooting Methods*, American Elsevier, New York, 1972.

An article which appeared too late to be included in the text compares a large number of techniques on a large number of initial-value problems is

T. E. Hull, W. H. Emright, B. M. Fellen and A. E. Sedgwick, "Comparing Numerical Methods for Ordinary Differential Equations," *SIAM J. Numer. Anal.* **9** (1972), 603–637.

A portion of their abstract is included because of the importance of their conclusions.

"Numerical methods for systems of first order ordinary differential equations are tested on a variety of initial value problems. The methods are compared primarily as to how well they can handle relatively routine integration steps under a variety of accuracy requirements, rather than how well they handle difficulties caused by discontinuities, stiffness, roundoff or getting started."

"According to criteria involving the number of function evaluations, overhead cost, and reliability, the best general-purpose method, if function evaluations are not costly, is one due to Bulirsch and Stoer. However, when function evaluations are relatively expensive, variable-order methods based on Adams formulas are best. The overhead costs are lower for the method of Bulirsch and Stoer, but the Adams methods require considerably fewer function evaluations. Krogh's implementation of a variable-order Adams method is the best of those tested, but one due to Gear is also very good. In general, Runge-Kutta methods are not competitive, but fourth or fifth order methods of this type are best for restricted classes of problems in which function evaluations are not very expensive and accuracy requirements are not very stringent."

III

INTRODUCTION TO INVARIANT IMBEDDING

1 Introduction

What is *invariant imbedding*? The answer to this question is both simple and complicated. Although the actual application of the imbedding is relatively straightforward, the exact form of the imbedding to be used normally must be determined for each new problem. Basically, the method involves generating a "family" of problems by means of a single parameter, where the basic properties of the system remain invariant under the generation of the family. The family then provides a means of advancing from one member, sometimes degenerate, to the solution of the original problem.

In the original applications of invariant imbedding, the generating parameter was the "size" of the system; for example, the length of the interval or the thickness of a slab. In the past few years we have seen that the principles of invariance can take on many different forms. This leads to problems where the imbedding parameter can be any one of several "physical" parameters of the system. Many problems of classical analysis can also be viewed as an "imbedding," where the imbedding is almost always either position in a fixed interval or time.

The method of invariant imbedding, as well as its close associate, dynamic programming, is one of the most criticized and yet one of the most misunderstood techniques to appear since Lebesque introduced his controversial new concept of the integral in 1906. Many critics claim that the method does not deserve to be given a specific and exotic name, since the imbedding techniques of classical analysis have never achieved such status and that invariant imbedding has never shown its clear superiority over classical methods.* The author disagrees quite vigorously with both of these claims. This is not to imply

---

* We shall call the classical methods those which use either position in a fixed interval or time as the independent parameters.

that we believe that the imbedding should replace the classical methods. On the contrary, a well-informed scientist should have at his disposal a number of "weapons" with which to attack a given problem. In fact, in many applications the two methods complement each other.

Although some of the concepts of invariance date back to the 1880's, to the paper by Stokes, the real birth of the method of the invariant imbedding, as far as we are concerned, started with the paper of Bellman and Kalaba in 1956. In fact, it was Bellman who coined the term "invariant imbedding."

We shall start our study of invariant imbedding with several traditional applications, the order of which has no real significance. The applications will be simple and we shall parallel the classical approach for contrast. This will not only clarify the meaning of the imbedding procedure, but also clearly indicate the types of problems where the imbedding procedure has distinct advantages over classical procedures.

2 Trajectory Analysis – Classical

Suppose we wish to determine the maximum altitude that a ball will attain if it is thrown straight up into the air with an initial velocity, v. In order to simplify the analysis, let us assume that the height the ball attains is small compared to the radius of the earth so that the force of gravity can be assumed to be constant and that there is no air resistance.

With these basic assumptions, the only force acting on the ball is the force of gravity and, hence, Newton's second law translates into the mathematical equation

$$\frac{d^2 y}{dt^2} = - g, \quad y(0) = 0, \quad y'(0) = v, \tag{III.2.1}$$

where $y(t)$ is the height of the ball at any time, t, and we have taken the mass of the ball to be unity.

If we integrate both sides in (III.2.1) and use the initial condition $y'(0) = v$, we obtain

$$\frac{dy}{dt} = v - gt. \tag{III.2.2}$$

Since the velocity is zero at the maximum height, the time at which the maximum height is attained is given by

$$t_{max} = v/g. \tag{III.2.3}$$

An expression for the height at any time, t, can be found by integrating both sides of (III.2.2) and using $y(0)=0$,

$$y(t) = vt - \tfrac{1}{2}gt^2.$$ (III.2.4)

By substituting (III.2.3) into (III.2.4), we find the maximum altitude

$$y(t_{\max}) = v^2/2_g.$$ (III.2.5)

Observe that, in order to solve for the maximum height, we must solve for the height at any time, t. In this simple problem this causes no particular difficulty, but consider how inefficient the process would be for more complicated problems. This is particularly true if we want the maximum altitude for several values of the initial velocity, v, since the above process must be repeated for each new v.

3   Trajectory Analysis – Invariant Imbedding

Suppose we do not particularly care to know the position of the ball at all times, but only the maximum altitude of the ball as a function of the initial velocity, v. We saw that in the classical analysis approach we would have to repeat the above process for every new v.

To circumvent this problem, let us define the function:

$$f(v) = \text{maximum altitude a ball will attain as a result of}$$
$$\text{initial velocity, } v.$$ (III.3.1)

Let a small time Δ expire. During this time interval, the ball travels upward a distance $\Delta v + 0(\Delta^2)$ (see Figure 3-1.), and the ball losses $g\Delta + 0(\Delta^2)$ velocity. We may now view this process as a multistage process and consider the problem of determining the additional altitude required to reach the maximum. From the definition of $f(v)$ and from Figure 3-1, we see that the additional altitude would be $f(v-g\Delta)$. Thus, we have the relation

$$f(v) = v\Delta + f(v - g\Delta) + 0(\Delta^2).$$ (III.3.2)

By expanding $f(v-g\Delta)$ in a Taylor series, we have

$$f(v) = v\Delta + f(v) - g\Delta f'(v) + 0(\Delta^2)$$ (III.3.3)

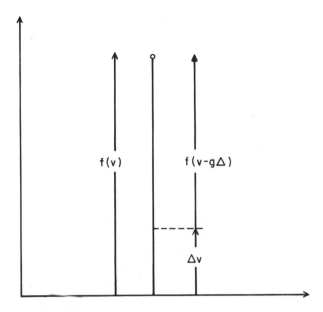

Figure 3-1.   Altitude as Function of Initial Velocity

If we divide both sides by $\varDelta$ and let $\varDelta \to 0$, we obtain the differential equation

$$f'(v) = v/g \qquad\qquad\qquad \text{(III.3.4)}$$

and the obvious initial condition $f(0) = 0$. The solution is

$$f(v) = v^2/2g, \qquad\qquad\qquad \text{(III.3.5)}$$

which agrees with the result of the previous section. However, once we have solved (III.3.4), we have obtained an expression for the maximum altitude for *any initial velocity*, v. Obviously, this formulation is much more efficient than the classical approach.

4   Trajectory Analysis – More Realistic Models

The model described above was simplified considerably. If we include such things as air resistance, the problem is somewhat complicated.

The classical equation can be written as

$$\frac{d^2y}{dt^2} = g\,(y,\,y'), \quad y\,(0) = 0, \quad y'\,(0) = v, \qquad \text{(III.4.1)}$$

where we have chosen to assume that the air resistance might be a function of both position and velocity. The procedure for analyzing this more realistic model in the classical sense is essentially the same as before. However, it may be necessary now to perform the integrations on a computer, and each new value of v definitely represents an entirely new procedure.

In order to treat this problem by invariant imbedding, we must observe that the maximum altitude attained is not only a function of the initial velocity but also a function of the initial height, h. Thus, we introduce the function

$$f\,(h,\,v) = \text{maximum altitude attained by a ball with initial}$$
$$\text{velocity, } v, \text{ and initial height, } h. \qquad \text{(III.4.2)}$$

As before, we start by letting a small time $\varDelta$ expire and then relate the height attained in the time $\varDelta$ to the overall maximum height. (See Figure 3-2.) Thus

$$f\,(h,\,v) = v\varDelta + f\,(h + v\varDelta,\,v + g\,(h,\,v)\,\varDelta) + 0\,(\varDelta^2). \qquad \text{(III.4.3)}$$

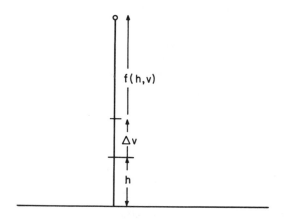

Figure 3-2.   Altitude as Function of Initial Velocity and Initial Height

Again, using a Taylor series expansion and letting $\Delta \to 0$, we obtain the partial differential equation

$$0 = v + v\,\frac{\partial f}{\partial h} + g\,(h,\,v)\,\frac{\partial f}{\partial v}, \tag{III.4.4}$$

with the initial condition

$$f\,(h,\,0) = h, \quad h \geqslant 0. \tag{III.4.5}$$

We must also use a numerical procedure, such as finite differences, the method of characteristics, or (III.4.3) itself, to solve (III.4.4, III.4.5). *The primary feature of the imbedding is that every value of $f\,(h,\,v)$ represents the solution of a specific problem.*

If the air resistance is such that (III.4.1) is a linear differential equation, then a new form of invariant imbedding, based upon a generalized Riccati transformation, simplifies the problem considerably. The new form of imbedding, which is the central theme of this book, also allows us to obtain the position of the ball at any time t. The older form of the imbedding is more versatile, but the new form has many very important features.

5 Particle Transport – Classical

In the previous sections, the classical and the invariant imbedding formulations were both initial-value problems. The principal advantage of the invariant imbedding approach was that it allowed certain types of parameter studies to be performed more efficiently than with the classical methods.

We now turn to problems for which the classical approach leads to two-point boundary-value problems. As we have seen in the introductory chapter, boundary-value problems are, in general, much more difficult to analyze than initial-value problems. We shall see, in the next section, that the invariant imbedding formulation remains an initial-value problem and, hence, we shall begin to see some of the real advantages of the invariant imbedding approach.

As our physical model we shall introduce the one-dimensional rod. It will

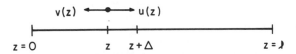

Figure 3-3.   One-Dimensional Rod Model for Neutron Transport

consist of an interval extending from $z=0$ to $z=l$, as in Figure 3-3. The rod will be capable of transporting particles, such as neutrons. The particles are allowed to travel only to the right or to the left and can interact only with the fixed constituents of the rod. When a particle interacts with the rod, the old particle disappears and two new ones appear, one traveling to the right and one traveling to the left. All the particles travel with the same speed and their other physical properties are such that the particles are distinguishable only by their direction. We shall consider only the steady-state case. The model is purposely chosen to be simple and not, necessarily, to approximate a true physical situation.

Let

$u(z)$ = expected density of particles at z and traveling to the right.

$v(z)$ = expected density of particles at z and traveling to the left.

$\sigma\Delta + o(\Delta)$ = probability of a collision occurring between a fixed nucleus and a particle moving between z and $z+\Delta$.

The collision probability is assumed to be independent of the direction of the moving particles; σ is called the *cross section* and, for the present, is assumed to be constant.

We now assume that we can determine, at a point, z, the flux or the number of particles per second moving to the left and to the right of the point. The number is also determined at $z+\Delta$, where $\Delta>0$.

The expected number of particles passing z to the right each second is $cu(z)$, where c is the particle speed in units per second. The expected number of particles suffering a collision in $(z, z+\Delta)$ is then

$$cu(z)\left[\sigma\Delta + o(\Delta)\right] = \sigma cu(z)\,\Delta + o(\Delta). \qquad \text{(III.5.1)}$$

From elementary probability theory we know that the expected number of particles emerging at $z+\Delta$ without experiencing a collision is given by

$$[1 - \sigma\Delta]\,cu(z) + o(\Delta). \qquad \text{(III.5.2)}$$

In addition, particles moving to the left which suffer collisions produce particles moving to the right. These number

$$\sigma cv(z + \Delta)\,\Delta + o(\Delta). \qquad \text{(III.5.3)}$$

Thus, at $z+\Delta$, we have

$$\begin{aligned}
cu(z + \Delta) &= \sigma cu(z)\,\Delta + (1 - \sigma\Delta)\,cu(z) + \sigma cv(z + \Delta)\,\Delta + o(\Delta) \\
&= cu(z) + \sigma cv(z + \Delta) + o(\Delta). \qquad \text{(III.5.4)}
\end{aligned}$$

We shall assume that $u(z)$ and $v(z)$ are continuous. Hence, we can divide through by $c\Delta$ and then take the limit as $\Delta \to 0$. Then (III.5.4) becomes

$$\frac{du}{dz} = \sigma v(z).$$

(III.5.5)

For the left-moving particles we obtain, in a similar fashion (see exercise 6),

$$-\frac{dv}{dz} = \sigma u(z).$$

(III.5.6)

For convenience, let us assume a set of simple boundary conditions. Let us suppose that, each second, a single particle is injected into the rod at $z = l$ and no particles enter at $z = 0$. Thus,

$$cu(0) = 0.$$

(III.5.7)

$$cv(l) = 1.$$

(III.5.8)

The solution of (III.5.5, III.5.6) with the above boundary conditions is

$$cu(z) = \sin \sigma z / \cos \sigma l$$

(III.5.9)

$$cv(z) = \cos \sigma z / \cos \sigma l.$$

(III.5.10)

This particular problem was designed to be quite easy to solve. However, if σ is allowed to be a function of z or if the interaction model is more complicated, it can be quite difficult to solve in closed form. As we have already seen, boundary-value problems usually require numerical techniques and can be quite tricky to solve. We shall return to this problem in Chapter IV.

6 Particle Transport – Invariant Imbedding

Suppose that we are not particularly interested in the internal flux, that is, we do not need to know $u(z)$ and $v(z)$ for all z, but only the number of particles being reflected, $u(l)$, or the number of particles being transmitted, $v(0)$, as a function of the length of the rod. This type of problem arises frequently in shielding problems of nuclear physics, electrochemical theory of fuel cells, and many other fields of mathematical physics.

In this section we shall derive differential equations for the unknown quantities $u(l)$ and $v(0)$ as a function of the length of the rod. The same model as before will be used except that, for esthetic reasons, the length of the rod will now be denoted by x. We define

  $cr(x)=$ the expected number of particles emerging to the right out of a rod
       of length x as a result of a unit input at that end.
  $ct(x)=$ the expected number of particles emerging to the left at $z=0$ out of
       a rod of length x as a result of a unit input at the right end.

The functions $r(x)$ and $t(x)$ are commonly known as the reflection and transmission functions, respectively, whereas $cr(x)$ and $ct(x)$ are known as the reflected flux and transmitted flux. From the above definitions, we observe that

$$r(x) = u(x) \tag{III.6.1}$$

$$t(x) = v(o). \tag{II.6.2}$$

Figure 3-4.   One-Dimensional Rod Model with Input from Right for Particle Counting

Consider the rod pictured in Figure 3-4. We wish to develop a relation between the reflection function for a rod of length x and the reflection function for a rod of length $x+\varDelta$. It is sometimes convenient to consider the rod of length x to be a *subrod* of the rod of length $x+\varDelta$.

As in the classical analysis we must consider the various interactions a particle can undergo while traversing the interval $(x, x+\varDelta)$. Let the particles enter the rod at $z=x+\varDelta$. While traversing the interval $(x, x+\varDelta)$, each particle has a probability of $\sigma\varDelta + o(\varDelta)$ of undergoing a collision. Each collision produces a particle traveling to the right and one traveling to the left. Since we do not count secondary collisions, the right-traveling particles exit at $x+\varDelta$. In addition, there is a probability of $(1-\sigma\varDelta)+o(\varDelta)$ that a particle entering at $x+\varDelta$ will traverse the interval $(x, x+\varDelta)$ without suffering a collision. Thus, we have

$$\sigma\varDelta + (1 - \sigma\varDelta) + o(\varDelta) = 1 + o(\varDelta) \tag{III.6.3}$$

particles traveling to the left at x; i.e., entering the subrod of length x. By definition this produces $cr(x)$ particles traveling to the right at x. Hereafter, we choose to ignore the $o(\varDelta)$ terms.

As these particles traverse the interval $(x, x+\varDelta)$, they will also be subject to collisions. Thus, every particle has a probability $\sigma\varDelta$ of collision, producing a right-moving and a left-moving particle, and a probability of $(1-\sigma\varDelta)$ of no interaction. Hence, we have $cr(x)$ particles which re-enter the subrod. Again, by definition, these produce $\sigma\varDelta c^2 r^2(x)$ particles emerging from the subrod at x and, hence, also at $x+\varDelta$. Thus, counting our contributions at $x+\varDelta$ (see Figure 3-5), we have

$$cr(x+\varDelta) = \sigma\varDelta + cr(x) + \sigma\varDelta c^2 r^2(x) + o(\varDelta). \qquad \text{(III.6.4)}$$

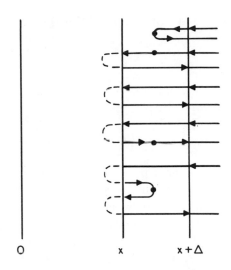

Figure 3-5.   Various Interactions that Particle May Undergo in Traversing Internal Length $\varDelta$

Dividing by $\varDelta$ and taking the limit $\varDelta \to 0$, we obtain the differential equation

$$cr'(x) = \sigma + \sigma c^2 r^2(x). \qquad \text{(III.6.5)}$$

Obviously, a rod of length zero reflects no particles, so

$$cr(0) = 0. \qquad \text{(III.6.6)}$$

The solution of (III.6.5, III.6.6) is

$$r(x) = \frac{1}{c}\tan \sigma x.$$
(III.6.7)

This differential equation is a special case of the Riccati differential equations discussed in section 7 of Chapter 1.

Now we must derive the differential equation for the transmission function $t(x)$. In the above analysis, we found that energy particles entering at $x+\Delta$ actually produced two contributions to the subrod; namely, $1+\sigma\Delta cr(x)$. By definition of the transmission function, we have

$$ct(x + \Delta) = (1 + \sigma\Delta cr(x))\, ct(x).$$
(III.6.8)

Again, by dividing by Δ and letting $\Delta \to 0$, (III.6.8) becomes

$$ct'(x) = \sigma c^2 r(x)\, t(x).$$
(III.6.9)

A rod of zero length with an input of one particle per second obviously transmits one particle per second. Hence,

$$ct(0) = 1.$$
(III.6.10)

There are several interesting aspects of the imbedding formulation. Notice that the unknown boundary condition, $u(x)=r(x)$, is the solution of a nonlinear differential equation; whereas with $r(x)$ having been determined, $t(x)$ is the solution of a linear differential equation. This nonlinearity is of no particular consequence, since it is of first order and both have their initial values given at the same point. In fact, as we shall see later, these equations have very desirable stability characteristics. Also, as in section 3, every value of $r(x)$ and $t(x)$ represents a solution to a particular problem. Although we have not discussed how to use invariant imbedding to obtain internal values, the imbedding still presents a significant improvement over the classical approach for this type of problem in efficiency and stability.

Figure 3-6.   One-Dimensional Rod Model with Input from Left for Particle Counting

Let us now consider a series of alterations of our model. Instead of having the input on the right, let the input be injected on the left (see Figure 3-6) and define

$c\varrho(x)$ = the expected number of particles emerging to the left at $z=0$ out of a rod of length x as a result of a unit input at $z=0$.

$c\tau(x)$ = the expected number of particles emerging to the right out of a rod of length x as a result of a unit input at $z=0$.

Using the "particle-counting" methods above, we can show that the functions $\varrho(x)$ and $\tau(x)$ (see exercise 8) satisfy

$$\varrho'(x) = \sigma c\tau(x)\, t(x), \quad \varrho(0) = 0, \qquad \text{(III.6.11)}$$

$$\tau'(x) = \sigma c\tau(x)\, r(x), \quad \tau(0) = 1. \qquad \text{(III.6.12)}$$

Hereafter, we shall adopt the following notation for the transport problems: a subscript i will denote a particle emerging on the ith end of the rod, where $i=r$ refers to the right and $i=l$ refers to the left. With this convention, the above functions become

$$r(x) = r_r(x), \qquad \text{(III.6.13)}$$

$$t(x) = t_l(x), \qquad \text{(III.6.14)}$$

$$p(x) = r_l(x), \qquad \text{(III.6.15)}$$

$$\tau(x) = t_r(x). \qquad \text{(III.6.16)}$$

Now consider a model with internal sources. That is a problem with

$cS_i(x)\, \Delta$ = rate of emission of particles moving in the ith direction in the interval $(x, x+\Delta)$.

$ce_i(x)$ = the expected number of particles emerging from the ith end of the rod of length x as a result of particles originating from the internal sources $S_r(z')$ and $S_l(z')$, $c \leqslant z' \leqslant x$.

The functions $e_i(x)$ are commonly known as the *escape* functions. The differential equations satisfied by $e_r(x)$ and $e_l(x)$ (see exercise 9) are

$$e_r'(x) = r_r(x)\, e_r(x) + S_l(x)\, e_r(x) + S_r(x), \quad e_r(0) = 0, \quad \text{(III.6.17)}$$

$$e_l'(x) = [e_r(x) + S_l(x)]\, t_r(x), \qquad\qquad\qquad e_l(0) = 0. \quad \text{(III.6.18)}$$

After having determined the six functions $r_i(x)$, $t_i(x)$, and $e_i(x)$, $i=l, r$, we are in a position to determine the internal flux from these quantities. This discussion will be deferred until Chapter IV.

7 Particle Transport – Riccati Transformation

Recall that the classical formulation of the simple rod model was represented by the system of two first order equations

$$\frac{du}{dz} = \sigma v(z), \qquad\qquad\qquad (III.7.1)$$

$$-\frac{dv}{dz} = \sigma u(z). \qquad\qquad\qquad (III.7.2)$$

By differentiating in (III.7.1) and using (III.7.2), we can rewrite the above system as

$$\frac{d^2 u}{dz} + \sigma^2 u(z) = 0, \qquad\qquad\qquad (III.7.3)$$

$$v(z) = \frac{1}{\sigma} u(z). \qquad\qquad\qquad (III.7.4)$$

Define the function $w(z)$ by the relation

$$\sigma u(z) = c w(z) u'(z). \qquad\qquad\qquad (III.7.5)$$

Differentiating in (III.7.5), we gct

$$\sigma u'(z) = c w(z) u'(z) + c w'(z) u'(z). \qquad\qquad\qquad (III.7.6)$$

If we use (III.7.3) and then (III.7.5) again, equation (III.7.6) can be rewritten as

$$u'(z) \{ c w'(z) - \sigma - \sigma c^2 w^2(z) \} = 0. \qquad\qquad\qquad (III.7.7)$$

We now argue that (III.7.7) must hold for all z and, hence, the term in braces must be zero. Thus,

$$c w'(z) = \sigma + \sigma c^2 w^2(z). \qquad\qquad\qquad (III.7.8)$$

From the relation (III.7.5), we see that

$$w(0) = 0, \qquad\qquad\qquad (III.7.9)$$

since, by our fundamental uniqueness theorem, $u'(0)=0$ would imply that $u(x) \equiv 0$.

Notice that (III.7.8, III.7.9) are identical to the invariant imbedding equations (III.6.5, III.6.6). Hence, at least in some cases, the invariant imbedding formulation and the Riccati transformation yield identical equations. It is this relationship between the two formulations that gives rise to the principal theme of the current study.

8 Particle Transport – Criticality

Recall that in section 5, the classical formulation led to the two-point boundary-value problem

$$u' = \sigma v \qquad (III.8.1)$$

$$- v' = \sigma u \qquad (III.8.2)$$

with

$$cu(0) = 0, \qquad cv(x) = 1, \qquad (III.8.3)$$

whose solution was

$$cu(z) = \sin \sigma z / \cos \sigma x \qquad (III.8.4)$$

$$cv(z) = \cos \sigma z / \cos \sigma x. \qquad (III.8.5)$$

Notice that if $x = (n + \frac{1}{2}) \pi / \sigma$ the denominators of (III.8.4, III.8.5) are zero and, hence, the solution of (III.8.1, III.8.3) is infinite. The first such x, in this case $x = \pi / 2\sigma$, is called the "critical" length or first "characteristic" length.

The differential equation for the reflection function of section 6 was given by

$$r'(x) = \sigma(1 + c^2 r^2(x)) \qquad (III.8.6)$$

with

$$r(0) = 0. \qquad (III.8.7)$$

Its solution is given by

$$cr(x) = \tan \sigma x = \sin \sigma x / \cos \sigma x. \qquad (III.8.8)$$

The reflection function also becomes infinite at $x = \pi / 2\sigma$. This is to be expected, inasmuch as $r(x) = u(x)$. However, the process of finding the critical length is much simplier from the reflection function, since we need only integrate (III.8.6, III.8.7) until it becomes infinite.

This discussion suggests that the method of invariant imbedding might be useful in solving eigenvalue problems. Actually, this is one of the most practical applications of invariant imbedding, a fact attested to by our devoting the entire Chapter V to such problems.

9 Exercises

1. Suppose we wish to determine, by using the invariant imbedding, the time required for the ball to attain its maximum altitude. Let $h(v)$ denote this time. Show that if Δ is an infinitesimal, $h(v) = \Delta + h(v - g\Delta)$ and thus $h'(v) = 1/g$. Using the initial condition $h(0) = 0$, show that $h(v) = v/g$.

   (Bellman and Kalaba)

2. A rabbit, initially at $(r, 0)$, travels along the x-axis in the direction of increasing x with constant velocity v_r. A dog, initially at $(0, d)$: follows the rabbit with constant velocity v_d, with the direction of motion pointed toward the rabbit at all times. Determine the path of the dog. (Hint: Since the dog points at the rabbit at all times, t, the tangent to the curve traced out by the dog's motion passes through the position of the rabbit at time, t).

   (Bellman and Kalaba)

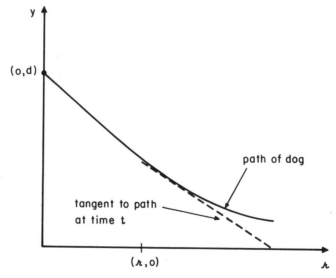

3. Let $f(r, d)$ denote the time at which the dog catches the rabbit. Show that

$$v_r \frac{\partial f}{\partial r} - \frac{dv_d}{\sqrt{r^2 + d^2}} \frac{\partial f}{\partial d} + 1 = 0, \quad f(r, 0) = \frac{r}{v_d - v_r}.$$

   (Bellman and Kalaba)

4. Let $g(r, d)$ denote the position along the x-axis where the dog catches the rabbit. Derive a partial differential equation for the quantity $g(r, d)$.

(Bellman and Kalaba)

5. Use the method of characteristics to find $f(r, d)$ and $g(r, d)$ in explicit analytic terms.

(Bellman and Kalaba)

6. Derive the differential equation (III.5.6) for v; i.e.,

$$-v' = \sigma u.$$

7. Show that the system (III.5.5–III.5.8) can be rewritten as

$$u''(z) + \sigma^2 u(z) = 0, \quad u(0) = 0, \quad u'(l) = \sigma/c,$$

$$v(z) = \frac{1}{\sigma} u'(z)$$

8. Derive the differential equations (III.6.11, III.6.12) for $r_l(z)$ and $t_r(z)$ by means of "particle-counting."

9. Derive the differential equations (III.6.17, III.6.18) for $e_r(z)$ and $e_l(z)$ by means of "particle-counting."

10. Derive the differential equations for the reflection and transmission functions for the model

11. The boundary conditions used thus far assume that once a particle leaves the rod it cannot return. Now suppose, for example, that a material is adjoined to the left end of the rod so that a certain portion of the particles which impinge upon it are returned to the rod. This type of problem is termed an *albedo* problem and, mathematically, we have

$$c\beta v(0) = cu(0).$$

The number β is called the albedo. Our previous boundary condition corresponds to $\beta=0$. (Until very recently it was thought that invariant imbedding was incapable of handling such a problem. We shall discuss one method here and another in Chapter IV.) Consider the system of two first-order linear equations

$$u'(z) = a(z)u(z) + b(z)v(z), \qquad \alpha_1 u(0) + \beta_1 v(0) = \gamma_1,$$
$$-v'(z) = c(z)u(z) + d(z)v(z), \qquad \alpha_2 u(x) + \beta_2 v(x) = \gamma_2.$$

Show that the transformation

$$\hat{u}(z) = \alpha_1 u(z) + \beta_1 v(z)$$
$$\hat{v}(z) = \alpha_2 u(z) + \beta_2 v(z)$$

converts the above problem into one of the same form but with

$$\hat{u}(0) = \gamma_1,$$
$$\hat{v}(x) = \gamma_2.$$

12. Show that the solutions of (III.6.5, III.6.6) and (III.6.9, III.6.10) satisfy the identity

$$1 + c^2 r^2 = c^2 t^2.$$

13. Show that if $u(z)$ satisfies

$$u''(z) + f(z)u'(z) + g(z)u(z) = 0,$$
$$u(0) = 0, \qquad u'(1) = 1,$$

and that if v satisfies

$$v''(z) + f(z)v'(z) + g(z)v(z) = 0,$$
$$v(0) = 0, \qquad v'(1) = \beta,$$

then

$$v(z) = \beta u(z).$$

14. All of our transport models were assumed to be linear. However, if we allow for particle-particle interactions to take place, a nonlinearity is introduced into the equations and some interesting results are obtained. Let us assume that in the interval $(z, z+\Delta)$ particles moving to the left

may collide with particles moving to the right, thereby resulting in an annihilation of the particles. (This is just one of the several physical results.) If we consider a process where the rate of annihilation of particles is proportional to the product of the right-moving and left-moving particles, the classical formulation can be written as

$$u' = v - \varepsilon uv,$$
$$-v' = u - \varepsilon uv, \qquad \varepsilon > 0$$
$$u(0) = 0, \qquad v(x) = y,$$

where we have taken $c = 1$, $\sigma = 1$. The input y is introduced because the simple boundary condition $v(x) = 1$ no longer suffices to give us the behavior of the system. We no longer have a simple relation as in exercise 13. The functions u and v are generally complicated functions of y.

Show that $u(z)$ is always an increasing function of z, and $0 \leqslant v(z) \leqslant \max(y, b^{-1})$. Thus, regardless of the value of x and of the input y, a critical situation can never be reached, however small b may be.

(Bellman, Kalaba and Wing)

10   Bibliographical Discussion

Section 1

The oldest known papers dealing with principles of invariance are:

G. G. Stokes, *Mathematical and Physical Papers*, Vol. 2, Cambridge University Press, 1880.
H. W. Schmidt, "Über Reflexion, und Absorption von β-Strahlen," *Annalen der Physik*, **23** (1907), 671–677.
V. A. Ambarzumian, "Diffuse Reflection of Light by a Foggy Medium," *Comptes Rendus* (Doklady) de l'Academie des Sciences de l'URSS, **38** (1943), 229–232.
S. Chandrasekhar, *Radiative Transfer*, Oxford, Clarendon Press, 1950.

The real birth of invariant imbedding as we know it today began with the following paper:

R. E. Bellman and R. E. Kalaba, "On the Principle of Invariant Imbedding and Propagation through Inhomogeneous Media," *Proc. Nat'l. Acad. Sci.*, U.S.A. **42** (1956), 629–632.

A long series of papers by Bellman, Kalaba, and G. M. Wing which dealt with neutron transport theory, radiative transfer, and wave propagation were then generated. For an extensive list see:

M. R. Scott, *A Bibliography on Invariant Imbedding and Related Topics*, Sandia Laboratories Report SC-B-71-0886, Albuquerque, New Mexico.

Section 2

The material in this section illustrates that the invariant imbedding (at least in this application) may be thought of as a decision-free dynamic programming process. See

R. E. Bellman, *Some Vistas of Modern Mathematics*, Univ. Kentucky Press, Lexington, 1968.

R. E. Bellman and R. E. Kalaba, *Dynamic Programming and Modern Control Theory*, Academic Press, New York, 1966.

Section 3

Much of the early work on invariant imbedding centered around its applications to one-dimensional particle transport. See

R. E. Bellman and R. E. Kalaba, "Transport Theory and Invariant Imbedding," *American Math. Soc.*, Providence, 1961.

For some applications of invariant imbedding to problems not covered in this book, see

E. Angel and R. E. Bellman, *Dynamic Programming and Partial Differential Equations*, Academic Press, New York, 1972.

P. B. Bailey and G. M. Wing, "Some Recent Developments in Invariant Imbedding with Applications," *J. Math. Phys.* **6** (1965), 453–462.

R. E. Bellman, R. E. Kalaba, and G. M. Wing, "Invariant Imbedding and Mathematical Physics I, Particle Processes," *J. Math. Phys.* **1** (1960), 280–308.

C. W. Maynard and M. R. Scott, "Invariant Imbedding and Linear Partial Differential Equations via Generalized Riccati Transformations," *J. Math. Anal. Appl.* **36** (1971), 432–459.

R. W. Preisendorfer, *Radiative Transfer on Discrete Spaces*, Pergamon Press, New York, 1965.

R. M. Redheffer, "On the Relation of Transmission-Line Theory to Scattering and Transfer," *J. Math. and Phys.* **41** (1962), 1.

IV

LINEAR BOUNDARY-VALUE PROBLEMS - INHOMOGENEOUS

1 Introduction

In Chapter III we showed how certain physical problems could be reformulated into more efficient procedures by using the method of invariant imbedding. However, we did not attempt to find any information about the position of the ball at any time, t, in the trajectory analysis case or any information about the internal flux in the transport model. *The inability to solve for internal values has always been one of the major criticisms of invariant imbedding.* In this chapter, we shall show that this criticism is not warranted. In fact, a number of procedures have been suggested for solving for the internal values. We shall develop and compare two such methods.

2 The Basic Invariant Imbedding Algorithm

Let us consider the linear boundary-value problem

$$\frac{\partial u(z, x)}{\partial z} = a(z)\, u(z, x) + b(z)\, v(z, x) + e(z), \qquad \text{(IV.2.1)}$$

$$-\frac{\partial v(z, x)}{\partial z} = c(z)\, u(z, x) + d(z)\, v(z, x) + f(z), \qquad \text{(IV.2.2)}$$

subject to

$$u(0, x) = \alpha, \qquad \text{(IV.2.3)}$$

$$v(x, x) = \beta. \qquad \text{(IV.2.4)}$$

We have written u and v as functions of two variables to emphasize that the solution is a function of the interval length, x.

We shall adopt the following "no nonsense" policy: *Unless specifically stated otherwise, the coefficients, $a(z)$, $b(z)$, $c(z)$, and $d(z)$; the forcing terms $e(z)$ and $f(z)$; the boundary conditions α and β; and the interval length x are such that the system (IV.2.1–IV.2.4) has a unique solution for all interval lengths $x \leqslant X$.* This basic restriction can be removed but only after we have discussed homogeneous problems in Chapter V.

The Riccati transformation, as described in Chapter I, is no longer valid because of the presence of the inhomogeneous terms in (IV.2.1–IV.2.3). In order to account for the inhomogeneities, we must introduce a new function $r_2(z)$ and the modified Riccati transformation.

$$u(z, x) = r_1(z) v(z, x) + r_2(z). \tag{IV.2.5}$$

Differentiating in (IV.2.5) with respect to z, we obtain

$$\frac{\partial u}{\partial z} = r_1'(z) v(z, x) + r_1(z) \frac{\partial v}{\partial z} + r_2'(z). \tag{IV.2.6}$$

If we now substitute (IV.2.1) and (IV.2.2) into (IV.2.6), then use (IV.2.5) again and simplify, we get

$$v(z, x) \{ b(z) + [a(z) + d(z)] r_1(z) + c(z) r_1^2(z) - r_1'(z) \}$$
$$+ \{ [a(z) + c(z) r_1(z)] r_2(z) + f(z) r_1(z) + e(z) - r_1^2(z) \} = 0. \tag{IV.2.7}$$

There are two ways to proceed from this point. We could argue either that (IV.2.7) must hold for all points z, whereby each term in braces must be zero, or that (IV.2.7) is satisfied if each term in braces is set to zero. In either case we shall prove that this procedure does, in fact, solve the original problem. Thus, we have the equations

$$r_1'(z) = b(z) + [a(z) + d(z)] r_1(z) + c(z) r_2(z), \tag{IV.2.8}$$

and

$$r_2'(z) = [a(z) + c(z) r_1(z)] r_2(z) + f(z) r_1(z) + e(z). \tag{IV.2.9}$$

A natural set of initial conditions for (IV.2.8, IV.2.9) may be obtained by evaluating (IV.2.5) at $z=0$,

$$u(0, x) = \alpha = r_1(0) v(0, x) + r_2(0).\qquad\text{(IV.2.10)}$$

The term $v(0, x)$, although unknown, is certainly finite. Hence, if we let

$$r_1(0) = 0,\qquad\text{(IV.2.11)}$$

then

$$r_2(0) = \alpha.\qquad\text{(IV.2.12)}$$

We could, of course, use other initial conditions for $r_1(0)$ and $r_2(0)$. However, if we define $r_1(z)$ as in Chapter III, then $r_1(0)=0$ is a physically meaningful choice, as is $r_2(0)=\alpha$, and r_1 and r_2 are independent of x. We shall return to a discussion of this in section 4.

Again we have several options available (see exercise 1). We prefer to proceed as follows: We define the functions $q_1(z)$ and $q_2(z)$ by means of the *recovery transformation*

$$v(0, x) = q_1(z) v(z, x) + q_2(z).\qquad\text{(IV.2.13)}$$

The differential equations for $q_1(z)$ and $q_2(z)$ are derived in the same fashion as for $r_1(z)$ and $r_2(z)$. We obtain

$$q_1'(z) = [d(z) + c(z) r_1(z)] q_1(z),\qquad\text{(IV.2.14)}$$

$$q_2'(z) = [f(z) + c(z) r_2(z)] q_1(z).\qquad\text{(IV.2.15)}$$

A suitable set of initial conditions for $q_1(z)$ and $q_2(z)$ may be obtained by evaluating (IV.2.13) at $z=0$ and setting

$$q_1(0) = 1,\qquad\text{(IV.2.16)}$$

$$q_2(0) = 0.\qquad\text{(IV.2.17)}$$

The basic algorithm may be described as follows: We integrate equations (IV.2.8, IV.2.9), subject to (IV.2.11, IV.2.12), and equations (IV.2.14, IV.2.15), subject to (IV.2.16, IV.2.17) from $z=0$ to $z=X$, where X is the largest value of x that is desired. *Notice that these four equations form an initial-value problem.* The functions $r_1(z)$, $r_2(z)$, $q_1(z)$ and $q_2(z)$ are not functions of x, since each satisfies an initial-value problem with the initial conditions being given at $z=0$, and the functions a, b, c, d, e and f do not depend upon x. (See section 4 for further discussion of this in terms of the rod model.) In order to obtain the values of $u(z, x)$ and $v(z, x)$, we pick a value of $x \leqslant X$, say x_1, and evaluate (IV.2.13) at $z=x$ to find $v(0, x)$, i.e.,

$$v(0, x_1) = q_1(x_1) v(x_1, x_1) + q_2(x_1)$$
$$= q_1(x_1) \beta + q_2(x_1). \qquad \text{(IV.2.18)}$$

We then use (IV.2.13) to find $v(z, x_1)$ for all $z \in [0, x_1]$.

$$v(z, x_1) = \frac{v(0, x_1) - q_2(z)}{q_1(z)},$$
$$= \frac{q_1(x_1)}{q_1(z)} \beta + \frac{q_2(x_1) - q_2(z)}{q_1(z)}. \qquad \text{(IV.2.19)}$$

Our basic assumptions of uniqueness prevent $q_1(z)$ from becoming zero over $0 \leqslant z \leqslant X$. Then $u(z, x_1)$ is obtained from

$$u(z, x_1) = r_1(z) v(z, x_1) + r_2(z). \qquad \text{(IV.2.20)}$$

If the solution is desired for another value of $x \leqslant X$, then *we do not have to perform any additional integrations.* We need only repeat the steps involving equations (IV.2.18–IV.2.20) with $x = x_2$. Since numerical integration is time consuming, this procedure offers a nice algorithm for varying the interval length. More important, the invariant imbedding equations tend to be more stable than (IV.2.1–IV.2.4).

Notice that if we wish to perform a parametric study with various values of $\beta = y(x, x)$, again we need only repeat the process (IV.2.18–IV.2.20). However, the process is somewhat more complicated to vary α. We shall eliminate this small difficulty in section 4.

3   Validation of the Basic Algorithm

We must prove that the basic algorithm introduced in the previous section does, indeed, solve the original boundary-value problem. In order to see this, we define new functions $\hat{u}(z, x)$ and $\hat{v}(z, x)$ by the relations

$$\hat{u}(z, x) = r_1(z) \hat{v}(z, x) + r_2(z) \qquad \text{(IV.3.1)}$$

and

$$\hat{v}(0, x) = q_1(z) \hat{v}(z, x) + q_2(z), \qquad \text{(IV.3.2)}$$

where $r_1(z)$, $r_2(z)$, $q_1(z)$ and $q_2(z)$ are defined by the differential equations

$$r_1'(z) = b(z) + [a(z) + d(z)] r_1(z) + c(z) r_1^2(z), \qquad r_1(0) = 0, \qquad \text{(IV.3.3)}$$

$$r_2'(z) = [a(z) + c(z)r_1(z)]r_2(z) + f(z)r_1(z) + e(z), \quad r_2(0) = \alpha,$$
$$\tag{IV.3.4}$$

$$q_1'(z) = [d(z) + c(z)r_1(z)]q_1(z), \quad q_1(0) = 1,$$
$$\tag{IV.3.5}$$

$$q_2'(z) = [f(z) + c(z)r_2(z)]q_1(z), \quad q_2(0) = 0.$$
$$\tag{IV.3.6}$$

Accepting this as our starting point, we differentiate in (IV.3.1) and (IV.3.2) with respect to z and then use (IV.3.3–IV.3.6). Upon simplifying, we find that $\hat{u}(z, x)$ and $\hat{v}(z, x)$ satisfy

$$\frac{\partial \hat{u}}{\partial z} = a(z)\hat{u}(z, x) + b(z)\hat{v}(z, x) + e(z), \tag{IV.3.7}$$

$$-\frac{\partial \hat{v}}{\partial z} = c(z)\hat{u}(z, x) + d(z)\hat{v}(z, x) + f(z). \tag{IV.3.8}$$

The condition

$$\hat{u}(0, x) = \alpha \tag{IV.3.9}$$

is easily verified by evaluating (IV.3.1) at $z=0$ and using the initial conditions on $r_1(z)$ and $r_2(z)$. To verify that $\hat{v}(x, x) = \beta$ is a little more complicated. Using (IV.3.1) we can write (IV.3.8) as

$$-\frac{\partial \hat{v}(z, x)}{\partial z} = [d(z) + c(z)r_1(z)]\hat{v}(z, x) + f(z) + c(z)r_2(z). \tag{IV.3.10}$$

Integrating both sides of (IV.3.10) from $z=0$ to $z=x$, we obtain

$$\hat{v}(0, x) = e^{\int_0^x [d(z)+c(z)r_1(z)]\,dz} \hat{v}(x, x)$$
$$+ \int_0^x e^{\int_0^z [d(s)+c(s)r_1(s)]\,ds} [f(z) + c(z)r_2(z)]\,dz. \tag{IV.3.11}$$

Now from (IV.3.2), evaluated at $z=x$, we have

$$\hat{v}(0, x) = q_1(x)\hat{v}(x, x) + q_2(x). \tag{IV.3.12}$$

Our validation procedure will be complete if we can identify

$$q_1(x) = e^{\int_0^x [d(z) + c(z) r_1(z)] \, dz} \qquad \text{(IV.3.13)}$$

and

$$q_2(x) = \int_0^x e^{\int_0^z [d(s) + c(s) r_1(s)] \, ds} \left[f(z) + c(z) r_2(z) \right] dz$$

$$\qquad \text{(IV.3.14)}$$

$$= \int_0^x q_1(z) \left[f(z) + c(z) r_2(z) \right] dz \,.$$

This is an easy identification, since integration of (IV.3.5) and (IV.3.6) gives precisely (IV.3.13) and (IV.3.14). Thus, we have completed the proof that our invariant imbedding procedure does indeed solve our original boundary-value problem.

4   An Analytical Example

In order to illustrate the foregoing procedures, we shall consider an example which can be carried to completion in closed form. We consider the linear equation

$$y'' = y + 1 \,, \qquad \text{(IV.4.1)}$$

subject to the boundary conditions

$$y(0) = \alpha \,, \quad y'(x) = \beta \,. \qquad \text{(IV.4.2)}$$

We first put (IV.4.1, IV.4.2) into the form of (IV.2.1–IV.2.4) by letting $u(z, x) = y(z)$ and $v(z, x) = y'(z)$. Then we have

$$u'(z, x) = v(z, x) \,, \qquad \text{(IV.4.3)}$$

$$- v'(z, x) = - u(z, x) - 1 \,, \qquad \text{(IV.4.4)}$$

subject to

$$u(0, x) = \alpha \,, \qquad \text{(IV.4.5)}$$

$$v(x, x) = \beta \,. \qquad \text{(IV.4.6)}$$

The differential equations for the invariant imbedding functions are

$$r_1'(z) = 1 - r_1^2(z), \qquad\qquad r_1(0) = 0, \qquad \text{(IV.4.7)}$$

$$r_2'(z) = -r_1(z)r_2(z) - r_1(z), \qquad r_2(0) = \alpha, \qquad \text{(IV.4.8)}$$

$$q_1'(z) = -r_1(z)q_1(z), \qquad\qquad q_1(0) = 1. \qquad \text{(IV.4.9)}$$

$$q_2'(z) = -[1 + r_2(z)]q_1(z), \qquad q_2(0) = 0. \qquad \text{(IV.4.10)}$$

The solutions, which are easily obtained, are

$$r_1(z) = \tanh z, \qquad\qquad\qquad\qquad\qquad \text{(IV.4.11)}$$

$$r_2(z) = (\alpha + 1)\operatorname{sech} z - 1, \qquad\qquad\qquad \text{(IV.4.12)}$$

$$q_1(z) = \operatorname{sech} z, \qquad\qquad\qquad\qquad\qquad \text{(IV.4.13)}$$

$$q_2(z) = -(\alpha + 1)\tanh z. \qquad\qquad\qquad\quad \text{(IV.4.14)}$$

The next step in the use of the algorithm is to evaluate (IV.2.13) at $z=x$. We obtain

$$v(0, x) = \beta \operatorname{sech} x - (\alpha + 1)\tanh x. \qquad\qquad \text{(IV.4.15)}$$

We now solve for $v(z, x)$ from (2.13) and obtain

$$v(z, x) = q_1^{-1}(z)\,[\beta q_1(x) + (q_2(x) - q_2(z))] \qquad\qquad \text{(IV.4.16)}$$

$$v(z, x) = \left[\frac{\beta - (\alpha + 1)\sinh x}{\cosh x}\right]\cosh z + (1 + \alpha)\sinh z. \qquad \text{(IV.4.17)}$$

The solution for $u(z, x)$ is found to be

$$u(z, x) = \left[\frac{\beta - (1 + \alpha)\sinh x}{\cosh x}\right]\sinh z + (1 + \alpha)\cosh z - 1.$$

$$\text{(IV.4.18)}$$

The reader can easily verify that $u(z, x) = y(z)$ is a solution of (IV.4.1) and that (IV.4.17) and (IV.4.18) satisfy the given boundary conditions.

5   Generalization of the Basic Algorithm

In certain applications it is desirable to obtain a more complete resolution of the modified Riccati transformation of section 2. Again we consider the system

$$\frac{\partial u\,(z,\,x)}{\partial z} = a\,(z)u\,(z,\,x) + b\,(z)v\,(z,\,x) + e\,(z), \qquad \text{(IV.5.1)}$$

$$-\frac{\partial v\,(z,\,x)}{\partial z} = c\,(z)u\,(z,\,x) + d\,(z)v\,(z,\,x) + f\,(z), \qquad \text{(IV.5.2)}$$

$$u\,(0,\,x) = \alpha, \qquad \text{(IV.5.3)}$$

$$v\,(x,\,x) = \beta. \qquad \text{(IV.5.4)}$$

We introduce the *generalized Riccati transformation*

$$u\,(z,\,x) = r_1\,(z)v\,(z,\,x) + r_2\,(z)u\,(0,\,x) + r_3\,(z) \qquad \text{(IV.5.5)}$$

and the corresponding *recovery transformation*

$$v\,(0,\,x) = q_1\,(z)v\,(z,\,x) + q_2\,(z)u\,(0,\,x) + q_3\,(z). \qquad \text{(IV.5.6)}$$

In section 9 we shall discuss the physical significance of each term as applied to the transport model.

The differential equations satisfied by the $r_i(z)$ and $q_i(z)$ $(i=1, 2, 3)$ functions may be derived exactly as before (see exercise 6):

$$r_1'\,(z) = b\,(z) + [a\,(z) + d\,(z)]\,r_1\,(z) + c\,(z)\,r_1^2\,(z), \qquad \text{(IV.5.7)}$$

$$r_2'\,(z) = [a\,(z) + c\,(z)r_1\,(z)]\,r_2\,(z), \qquad \text{(IV.5.8)}$$

$$r_3'\,(z) = [a\,(z) + c\,(z)r_1\,(z)]\,r_3\,(z) + f(z)r_1\,(z) + e\,(z), \qquad \text{(IV.5.9)}$$

$$q_1'\,(z) = [d\,(z) + c\,(z)r_1\,(z)]\,q_1\,(z), \qquad \text{(IV.5.10)}$$

$$q_2'\,(z) = c\,(z)q_1\,(z)r_2\,(z), \qquad \text{(IV.5.11)}$$

$$q_3'\,(z) = [c\,(z)r_3\,(z) + f(z)]\,q_1\,(z). \qquad \text{(IV.5.12)}$$

A suitable set of initial conditions for (IV.5.7–IV.5.12) is

$$\begin{array}{lll} r_1\,(0) = 0, & r_2\,(0) = 1, & r_3\,(0) = 0, \\ q_1\,(0) = 1, & q_2\,(0) = 0, & q_3\,(0) = 0. \end{array} \qquad \text{(IV.5.13)}$$

The method of solution is basically the same as for the basic algorithm as described in section 2 except that we now have six first-order equations to integrate and each is independent of the boundary conditions of (IV.5.1–IV.5.4). This allows for easy parametric study of the boundary conditions, as well as of the interval length, x. More important, the new transformations have a more meaningful physical interpretation.

In terms of the above functions the solutions for $u(z)$ and $v(z)$ become

$$v(z, x) = q_1^{-1}(z) \{q_1(x)\beta + \alpha[q_2(x) - q_2(z)] + [q_3(x) - q_3(z)]\}$$

(IV.5.14)

$$u(z, x) = r_1(z)v(z, x) + \alpha r_2(z) + r_3(z).$$

(IV.5.15)

6 General Boundary Conditions

Suppose we have the problem (IV.5.1, IV.5.2) subject to the boundary conditions

$$\alpha_1 u(0, x) + \beta_1 v(0, x) + \gamma_1 u(x, x) + \delta_1 v(x, x) = n_1,$$ (IV.6.1)

$$\alpha_2 u(0, x) + \beta_2 v(0, x) + \gamma_2 u(x, x) + \delta_2 v(x, x) = n_2.$$ (IV.6.2)

These very general boundary conditions are easily handled by using the generalized procedure of section 5. We evaluate (IV.5.5) and (IV.5.6) at $z = x$ and use the above boundary conditions to form a system of four equations for the four unknowns $u(0, x)$, $v(0, x)$, $u(x, x)$ and $v(x, x)$; i.e.,

$$\begin{bmatrix} \alpha_1 & \beta_1 & \gamma_1 & \delta_1 \\ \alpha_2 & \beta_2 & \gamma_2 & \delta_2 \\ r_2(x) & 0 & -1 & r_1(x) \\ q_2(x) & -1 & 0 & q_1(x) \end{bmatrix} \begin{bmatrix} u(0, x) \\ v(0, x) \\ u(x, x) \\ v(x, x) \end{bmatrix} \begin{bmatrix} n_1 \\ n_2 \\ -r_3(x) \\ -q_3(x) \end{bmatrix}$$ (IV.6.3)

Once this system has been solved, the internal values, $u(z, x)$ and $v(z, x)$, for $0 \leqslant z \leqslant x$, may be obtained by solving (IV.5.5) and (IV.5.6).

7 Inverse Riccati Transformations

In certain problems we may desire to write our Riccati and recovery transformations in the form

$$v(z, x) = s_1(z)u(z, x) + s_2(z)v(0, x) + s_3(z)$$ (IV.7.1)

$$u(0, x) = t_1(z)u(z, x) + t_2(z)v(0, x) + t_3(z), \qquad \text{(IV.7.2)}$$

which are, in some sense, inverse transformations to those introduced in the previous section. The differential equations satisfied by the s and t functions are

$$-s_1'(z) = c(z) + [a(z) + d(z)]s_1(z) + b(z)s_1^2(z), \qquad \text{(IV.7.3)}$$

$$-s_2'(z) = [d(z) + b(z)s_1(z)]s_2(z), \qquad \text{(IV.7.4)}$$

$$-s_3'(z) = [d(z) + b(z)s_1(z)]s_3(z) + e(z)s_1(z) + f(z), \qquad \text{(IV.7.5)}$$

$$-t_1'(z) = [a(z) + b(z)s_1(z)]t_1(z), \qquad \text{(IV.7.6)}$$

$$-t_2'(z) = b(z)t_1(z)s_2(z), \qquad \text{(IV.7.7)}$$

$$-t_3'(z) = [b(z)s_3(z) + e(z)]t_1(z), \qquad \text{(IV.7.8)}$$

with the initial conditions

$$\begin{array}{lll} s_1(0) = 0, & s_2(0) = 1, & s_3(0) = 0, \\ t_1(0) = 1, & t_2(0) = 0, & t_3(0) = 0. \end{array} \qquad \text{(IV.7.9)}$$

The new transformations defined by (IV.7.1, IV.7.2) are sometimes used for several reasons: The form of the given boundary or initial conditions may be better suited for these transformations, depending upon the coefficients a, b, c, or d; the differential equations (IV.7.3–IV.7.8) may be more stable; and these transformations are more natural to generalize for partial differential equations. Modified versions of both sets of transformations (IV.5.5, IV. 5.6)and (IV.7.1, IV.7.2) are used to compute characteristic values and characteristic lengths for ordinary differential equations.

8   Relationship to Green's Functions

In many applications of mathematical physics, Green's functions play a fundamental role in the analytical study of various linear boundary-value problems. However, Green's functions have seldom been used in the numerical solution of such problems, chiefly because of a lack of adequate numerical procedures for generating these functions.

In this section, we shall develop several initial-value procedures for generating Green's functions which are of interest both analytically and numerically. Consider the second-order linear equation

$$y''(z) + a(z)y'(z) + b(z)y(z) = -h(z), \qquad \text{(IV.8.1)}$$

subject to the two-point boundary conditions

$$y(0) = 0, \quad y'(x) = 0. \tag{IV.8.2}$$

The terms r_3 and q_3 of section 5 were included to account for the source term $h(z)$ and, every time a new source term is considered, the differential equations for r_3 and q_3 must be re-integrated. If a large number of source terms are considered, a lot of computing time is consumed. In order to circumvent this, we rewrite (IV.5.5, IV.5.6) as

$$y(z, x) = r_1(z) y'(z, x) + \int_0^z r_4(z, z') h(z') \, dz', \tag{IV.8.3}$$

$$y'(0, x) = q_1(z) y'(z, x) + \int_0^z q_4(z, z') h(z') \, dz', \tag{IV.8.4}$$

where r_1 and q_1 are the same as in (IV.5.5, IV.5.6) and r_2 and q_2 are not required since $y(0, x) = 0$.

If we differentiate in (IV.8.3) with respect to z, use (IV.8.1) and (IV.8.3) to eliminate $y(z, x)$, and simplify, we obtain

$$0 = \{r_1'(z) - 1 - a(z) r_1(z) - b(z) r_1^2(z)\}$$

$$+ \left\{ - \int_0^z b(z) r_1(z) r_4(z, z') h(z') \, dz' + [r_4(z, z') - r_1(z)] h(z) \right.$$

$$+ \left. \int_0^z r_4'(z, z') h(z') \, dz' \right\}. \tag{IV.8.5}$$

We next introduce the delta function into the second term in braces and obtain

$$0 = \{r_1'(z) - 1 - a(z) r_1(z) - b(z) r_1^2(z)\}$$

$$+ \left\{ \int_0^z [r_4'(z, z') - b(z) r_1(z) r_4(z, z') \right.$$

$$+ \left. (r_4(z, z') - r_1(z')) \delta(z' - z)] h(z') \, dz' \right\}. \tag{IV.8.6}$$

We now argue that the above equation must hold for all z, and this implies that each term in braces is zero. In addition, since the second term in braces is zero and must hold for all source terms $h(z)$, the term in brackets must also be zero. Thus, we have

$$r_1'(z) = 1 + a(z)r_1(z) + b(z)r_1^2(z),$$ (IV.8.7)

$$r_4'(z, z') = b(z)r_1(z)r_4(z, z') + [r_1(z) - r_4(z, z')]\delta(z' - z).$$ (IV.8.8)

The initial conditions for (IV.8.7) and (IV.8.8) are given by

$$r_1(0) = 0,$$ (IV.8.9)

$$r_4(0, z') = 0.$$ (IV.8.10)

However, for $z < z'$, (IV.8.8) is a homogeneous equation with homogeneous initial condition. Thus, $r_4(z, z') = 0$ for $z < z'$. At $z = z'$, we must have $r_4(z, z') = r_1(z')$. We can now dispense with the delta function in (IV.8.8) and rewrite it as

$$r_4'(z, z') = b(z)r_1(z)r_4(z, z'), \quad z' < z,$$ (IV.8.11)

subject to the initial condition

$$r_4(z', z') = r_1(z').$$ (IV.8.12)

Using similar techniques, we find that the differential equations for $q_1(z)$ and $q_4(z, z')$ are

$$q_1'(z) = [a(z) + b(z)r_1(z)]q_1(z),$$ (IV.8.13)

$$q_4'(z, z') = b(z)q_1(z)q_4(z, z'),$$ (IV.8.14)

and their initial conditions are

$$q_1(0) = 1$$ (IV.8.15)

$$q_4(z', z') = q_1(z').$$ (IV.8.16)

The procedure for generating $y(z, x)$ and $y'(z, x)$ is now very similar to those of previous sections. The differential equations for r_1, q_1, r_4 and q_4 are integrated over $0 \leqslant z \leqslant X$. The equations for r_4 and q_4 are integrated for various values of z'. We then pick a desired interval length, x, and evaluate (IV.8.4) at $z = x$ to obtain

$$y'(0, x) = \int_0^x q_4(x, z')h(z')\,dz'.$$ (IV.8.17)

Then (IV.8.3) and (IV.8.4) can be rewritten as

$$y'(z, x) = q_1^{-1}(z)\left\{\int_0^x q_4(x, z') h(z') dz' - \int_0^z q_4(z, z') h(z') dz'\right\} \qquad \text{(IV.8.18)}$$

$$y(z, x) = r_1(z) y'(z, x) + r_4(z, z'). \qquad \text{(IV.8.19)}$$

This represents the solution of (IV.8.1, IV.8.2) for a particular value of x and a particular source term $h(z)$. In contrast to the previous procedures, we do not need to re-integrate in order to change the interval length x or the source term $h(z)$. We need only repeat steps (IV.8.17–IV.8.19) for a new $x \leqslant X$ or a new source term $h(z)$.

In view of the form of (IV.8.3, IV.8.4) a natural question arises: Is there a relationship between the invariant imbedding equations and the Green's functions? Not only is the answer in the affirmative, but the relationship is useful from both an analytical and numerical viewpoint.

Let $g(z, z', x)$ denote the Green's function for the operator

$$Ly = y''(z) + a(z) y'(z) + b(z) y(z) = - h(z) \qquad \text{(IV.8.20)}$$

with the boundary conditions

$$y(0, x) = 0, \quad y'(x, x) = 0. \qquad \text{(IV.8.21)}$$

Then the solution of (IV.8.20, IV.8.21) can be represented as

$$y(z, x) = \int_0^x g(z, z', x) h(z') dz'. \qquad \text{(IV.8.22)}$$

Substitution of (IV.8.22) into (IV.8.3) yields

$$\int_0^x g(z, z', x) h(z') dz' = r_1(z) \int_0^x g'(z, z', x) h(z') dz' + \int_0^z r_4(z, z') h(z') dz'. \qquad \text{(IV.8.23)}$$

In order to have all integrals over the same interval, we introduce the Heaviside step function into the last integral on the right. Then (IV.8.23) becomes

$$0 = \int_0^x \{g(z, z', x) - r_1(z) g'(z, z', x) - r_4(z, z') [1 - H(z' - z)]\} h(z') dz' \qquad \text{(IV.8.24)}$$

Again we argue that the term in the braces must be zero, since the integral must be zero for all source terms, $h(z)$. Thus, we have

$$r_4(z, z')\,[1 - H(z' - z)] = g(z, z', x) - r_1(z)g'(z, z', x), \quad \text{(IV.8.25)}$$

or

$$r_4(z, z') = \begin{cases} 0, & z < z' \\ g(z, z', x) - r_1(z)g'(z, z', x), & z' < z. \end{cases} \quad \text{(IV.8.26)}$$

Notice that, although the Green's function is symmetric in z and z' and is a function of the interval length, x, the function $r_4(z, z')$ is not symmetric and is not a function of the interval length, x.

The exact numerical procedure for generating the Green's function usually depends on the boundary conditions of the original problem. For the boundary conditions given in (IV.8.21), we propose the following procedure. Since $y'(x)=0$, we have $g'(x, z, x)=0$. From (IV.8.26) this implies that

$$g(x, z', x) = r_4(x, z'). \quad \text{(IV.8.27)}$$

Hence, we integrate (IV.8.7), subject to (IV.8.9), and (IV.8.11), subject to (IV.8.12), from $z=0$ to $z=X$. For the desired interval length, x, we integrate (IV.8.26), subject to (IV.8.27), from $z=x$ to $z=0$ for various values of z'.

9 Particle Transport – Revisited

We shall return to the transport model of Chapter III with a few modifications. We wish to consider the steady-state transport of monoenergetic particles of speed, c, in a one-dimensional rod. We introduce the optical depth, z, as the independent variable, so that the total macroscopic cross section is unity. The left end of all rods considered will be located at $z=0$. In accordance with the fundamental viewpoint of invariant imbedding, we consider a family of rods with right ends located at $z=x$, where x is a nonnegative number. Each rod is assumed to have the following properties:

   (i) The particles are allowed to travel only to the left or the right; they are not allowed to interact with each other but, rather, with the fixed nuclei of the rod.

  (ii) $a_{ij}(z)\,\varDelta =$ the *net* expected number of particles traveling in the jth direction which are produced by collision of a particle traveling in the ith direction while traversing the interval $(z, z+\varDelta)$, where $i=r$ refers to the right and $i=l$ refers to the left.

(iii) $cS_i(z)\Delta$ = rate of emission of particles moving in the ith direction in the interval $(z, z+\Delta)$.

We assume that the above functions are nonnegative and are well defined for $0 \leqslant z \leqslant X$, where X is the largest rod length of interest. We now introduce the functions $u(z, x)$ and $v(z, x)$, which are, respectively, the densities of particles traveling to the right and to the left at the point z in a rod of length, x, with $0 \leqslant z \leqslant x$. Our basic problem is to study the functions $u(z, x)$, $v(z, x)$ as a function of z and the length of the rod, x.

Classical conservation considerations of the above physical process yields the linear two-point boundary-value problem (sometimes referred to as the Boltzmann formulation)

$$\frac{\partial u(z, x)}{\partial z} = a_{rr}(z)u(z, x) + a_{lr}(z)v(z, x) + S_r(z), \qquad \text{(IV.9.1)}$$

$$-\frac{\partial v(z, x)}{\partial z} = a_{rl}(z)u(z, x) + a_{ll}(z)v(z, x) + S_l(z), \qquad \text{(IV.9.2)}$$

with various boundary conditions.

Let us now define a secondary problem by considering that portion of the rod extending from 0 to z; that is, we have a subrod of the given rod. In the following list of definitions, the terms "input" refer to number densities; for examples unity input means an input flux of c particles per second, where c = velocity.

Let

(i) $r_r(z)$ be the output of right-moving particles at z as a result of a unit input of left-moving particles at z,

(ii) $t_r(z)$ be the output of left-moving particles at 0, as a result of a unit input of left-moving particles at z, and

(iii) $e_r(z)$ be the ouput of right-moving particles at z as a result of particles originating from the internal sources $S_r(z')$ and $S_l(z')$, $0 \leqslant z' \leqslant z$.

Also, let $r_l(z)$, $t_l(z)$ and $e_l(z)$ be defined, respectively, as were $r_r(z)$, $t_r(z)$ and $e_r(z)$, except with right and left interchanged, and with z and 0 interchanged. Thus, we have labeled the reflection and transmission functions, r and t, by subscripts indicating the input end and the escape functions e, are labeled by the output end.

At this point our analysis could proceed in two different ways. We could use the particle-counting method to derive the differential equations satisfied by the above functions and then form the relations

$$u(z, x) = r_r(z)v(z, x) + t_l(z)u(0, x) + e_r(z), \qquad \text{(IV.9.3)}$$

$$v(0, x) = t_r(z)v(z, x) + r_l(z)u(0, x) + e_l(z). \qquad \text{(IV.9.4)}$$

These relations simply represent the superposition of various contributions, respectively, to right-moving particles out the right end of the subrod of length z and to left-moving particles out the left end of the rod of length z. Alternatively, we could form the relations (IV.9.3, IV.9.4) and use the analysis of the previous sections of this chapter to derive the differential equations satisfied by the reflection, transmission, and escape functions.

These differential equations are

$$r'_r(z) = a_{lr}(z) + [a_{rr}(z) + a_{ll}(z)] r_r(z) + a_{rl}(z) r_r^2(z), \qquad \text{(IV.9.5)}$$

$$t'_r(z) = [a_{ll}(z) + a_{rl}(z) r_r(z)] t_r(z), \qquad \text{(IV.9.6)}$$

$$e'_r(z) = [a_{rr}(z) + a_{rl}(z) r_r(z)] e_r(z) + S_l(z) r_r(z) + S_r(z), \qquad \text{(IV.9.7)}$$

$$t'_l(z) = [a_{rr}(z) + a_{rl}(z) r_r(z)] t_l(z), \qquad \text{(IV.9.8)}$$

$$r'_l(z) = a_{rl}(z) t_r(z) t_l(z), \qquad \text{(IV.9.9)}$$

$$e'_l(z) = [a_{rl}(z) e_r(z) + S_l(z)] t_r(z). \qquad \text{(IV.9.10)}$$

The initial conditions are obtained by considering a rod of zero length. Such a rod reflects no particles and cannot produce particles from internal sources. A rod of zero length does, however, transmit particles. Hence, we have the initial conditions

$$\begin{aligned} r_r(0) &= 0, & t_r(0) &= 1, & e_r(0) &= 0, \\ r_l(0) &= 0, & t_l(0) &= 1, & e_l(0) &= 0. \end{aligned} \qquad \text{(IV.9.11)}$$

It should be noted that the system (IV.9.5–IV.9.11) is essentially a triangular system; that is, if the integration is carried out in the order that the equations are written, the only unknown dependent variable appearing in each equation is the differential variable. In this sense, the system (IV.9.5–IV.9.11) is not strongly coupled.

10   The Method of Kagiwada and Kalaba

In this section we shall discuss the application of the classical form of invariant imbedding to boundary-value problems as introduced by H. H. Kagiwada, R. E. Kalaba, and their coworkers. The method to be described here is, of course, related to our previous discussions, but there are a few important differences.

We shall begin by considering the system

$$u_1(z, x) = a(z)u(z, x) + b(z)v(z, x), \qquad \text{(IV.10.1)}$$

$$- v_1(z, x) = c(z)u(z, x) + d(z)v(z, x), \qquad \text{(IV.10.2)}$$

subject to

$$u(0, x) = 0, \qquad \text{(IV.10.3)}$$

$$v(x, x) = 1, \qquad \text{(IV.10.4)}$$

where the notation $u_i(z, x)$, $i = 1, 2$, is used to denote differentiation with respect to the ith variable. We define

$$r(x) = u(z, x)\big|_{z=x} = u(x, x). \qquad \text{(IV.10.5)}$$

If we differentiate in (IV.10.5), we obtain

$$r'(x) = u_1(x, x) + u_2(x, x). \qquad \text{(IV.10.6)}$$

As before, we would like to express the right hand side of (IV.10.6) in terms of $r(x)$ and other known functions. A relation for $u_1(x, x)$ can easily be obtained by evaluating (IV.10.1) at $z = x$, followed by an application of (IV.10.4) and (IV.10.5). Hence

$$u_1(x, x) = a(x)r(z) + b(x). \qquad \text{(IV.10.7)}$$

We must now obtain a relation for $u_2(x, x)$. In order to do this, we first differentiate (IV.10.1–IV.10.4) with respect to x and find that $u_2(z, x)$ and $v_2(z, x)$ satisfy the system

$$u_{12}(z, x) = a(z)u_2(z, x) + b(z)v_2(z, x) \qquad \text{(IV.10.8)}$$

$$- v_{12}(z, x) = c(z)u_2(z, x) + d(z)v_2(z, x) \qquad \text{(IV.10.9)}$$

subject to

$$u_2(0, x) = 0 \qquad \text{(IV.10.10)}$$

$$v_2(x, x) = - v_1(x, x). \qquad \text{(IV.10.11)}$$

Observe that $u_2(z, x)$ and $v_2(z, x)$ satisfy the same differential equations as do $u(z, x)$ and $v(z, x)$ but with slightly different boundary conditions. Since the differential equations are linear, we can use exercise 13 of Chapter III to discover that $u_2(z, x)$ and $u(z, x)$ and $v_2(z, x)$ and $v(z, x)$ are merely multiples of each other; namely,

$$u_2(z, x) = - v_1(x, x) u(z, x), \qquad (IV.10.12)$$

$$v_2(z, x) = - v_1(x, x) v(z, x). \qquad (IV.10.13)$$

Evaluating (IV.10.2) at $z=x$ and using (IV.10.4) and (IV.10.5), we find that

$$- v_1(x, x) = c(x) r(x) + d(x). \qquad (IV.10.14)$$

Hence, we can rewrite (IV.10.12, IV.10.13) as

$$u_2(z, x) = [c(x) r(x) + d(x)] u(z, x), \qquad (IV.10.15)$$

$$v_2(z, x) = [c(x) r(x) + d(x)] v(z, x). \qquad (IV.10.16)$$

We are now in a position to find our relation for $u_2(x, x)$. We need only eva-luate (IV.10.15) at $z=x$. Hence,

$$u_2(x, x) = [c(x) r(x) + d(x)] r(x). \qquad (IV.10.17)$$

Substituting (IV.10.7) and (IV.10.17) into (IV.10.6), we find that $r(x)$ satisfies the differential equation

$$r'(x) = b(x) + [a(x) + d(x)] r(x) + c(x) r^2(x), \qquad (IV.10.18)$$

and the initial condition

$$r(0) \equiv 0 \qquad (IV.10.19)$$

can be found from (IV.10.3) and (IV.10.5).

In order to calculate the internal values (i.e., $u(z, x)$ for $0 < z < x$), we define

$$J(z, x) \equiv u(z, x) \qquad (IV.10.20)$$

with z fixed and x as the variable. Then, from (IV.10.15), it is easy to see that $J(z, x)$ satisfies the differential equation

$$J_2(z, x) = [c(x) r(x) + d(x)] J(z, x) \qquad (IV.10.21)$$

and, from (IV.10.5), has the initial condition

$$J(z, z) = r(z). \qquad (IV.10.22)$$

The internal values for $v(z, x)$ can be obtained in two different ways. We can define

$$K(z, x) \equiv v(z, x),\qquad\text{(IV.10.23)}$$

and then from (IV.10.16), $K(z, x)$ satisfies the linear differential equation

$$K_2(z, x) = [c(x)r(x) + d(x)]K(z, x)\qquad\text{(IV.10.24)}$$

with the initial condition, obtained from (IV.10.4) and (IV.10.23),

$$K(z, z) = 1.\qquad\text{(IV.10.25)}$$

Alternatively, since (IV.10.1–IV.10.4) is linear, we need only integrate either (IV.10.21, IV.10.22) or (IV.10.24, IV.10.25) and the other quantity may be obtained from the relation

$$u(z) = r(z)v(z).\qquad\text{(IV.10.26)}$$

Suppose we wish to calculate the values $u(0.2, x)$, $u(0.4, x), \ldots v(0.2, x)$, $v(0.4, x), \ldots$. The algorithm may be summarized as follows: We integrate

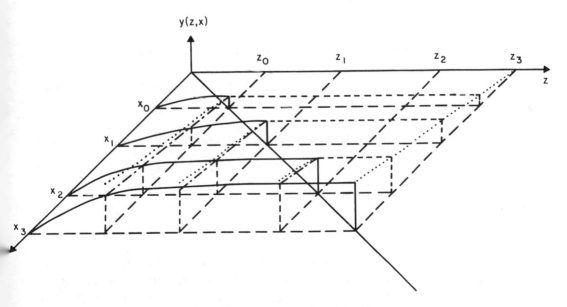

Figure 4-1.   Schematic of the Kagiwada and Kalaba Process

(IV.10.18, IV.10.19) from $x=0$ to $x=0.2$. At this point we adjoin (IV.10.24) with the initial condition $K(0.2, 0.2)=1$. We then integrate these two equations from $x=0.2$ to $x=0.4$ and adjoin another equation of the form (IV.10.24) with the initial condition $K(0.4, 0.4)=1$. This process is continued until we reach the maximum interval desired, and we have

$$u(0.2, x) = K(0.2, x),$$
$$u(0.4, x) = K(0.4, x),$$
$$\vdots \qquad ,$$

where each value of $K(z, x)$ represents an internal value for a fixed value of z and different interval lengths, x. The internal values for $u(z, x)$ are then obtained from (IV.10.26).

Perhaps we can best understand the above algorithm by viewing Figure 4–1. The classical solutions, for four different interval lengths x_0, x_1, x_2, x_3 are graphed as solid lines for fixed interval length, x, and variable z. The dash-dotted lines represent the values of $r(x)$. We need only integrate this equation once, since it is independent of z. However, for clarity, we have shown the solution for each fixed z. The dotted line represents the solution $u(z_i, x)$. Each value on a dotted line represents an internal value of $u(z, x)$ for a fixed value of z, but with different values of the interval length x. This is reinforced by the intersection of the dotted and solid lines.

11   The Method of Kagiwada and Kalaba – Source Terms

We shall now generalize the method of the previous section to handle problems with source or forcing terms. Consider the system

$$\hat{u}_1(z, x) = a(z)\hat{u}(z, x) + b(z)\hat{v}(z, x) + e(z), \qquad \text{(IV.11.1)}$$

$$-\hat{v}_1(z, x) = c(z)\hat{u}(z, x) + d(z)\hat{v}(z, x) + f(z), \qquad \text{(IV.11.2)}$$

subject to

$$\hat{u}(0, x) = 0 \qquad \text{(IV.11.3)}$$

$$\hat{v}(x, x) = \alpha. \qquad \text{(IV.11.4)}$$

The notation $\hat{u}(z, x)$ and $\hat{v}(z, x)$ will be used to denote the solution of the problem with source terms; namely, (IV.11.1–IV.11.4) while $u(z, x)$ and $v(z, x)$ will continue to denote the solution without source terms; namely (IV.10.1–IV.10.4).

We must modify our function in (IV.10.5) slightly, as we did in section 2, to account for the source terms. Hence, we introduce the relation

$$\hat{u}(x, x) = r(x)\alpha + q(x). \tag{IV.11.5}$$

If we differentiate in (IV.11.5), we obtain

$$\hat{u}_1(x, x) + \hat{u}_2(x, x) = r'(x)\alpha + q'(x). \tag{IV.11.6}$$

Remember that our purpose is to replace the terms on the left-hand side of (IV.11.6) with terms involving $r(x)$ and $q(x)$ and known terms. We can obtain a relation for $\hat{u}_1(x, x)$ by evaluating (IV.11.1) followed by the use of (IV.11.4) and (IV.11.5) Thus, we have

$$[a(x)r(x) + b(x)]\alpha + a(x)q(x) + e(x) + \hat{u}_2(x, x) = r'(x)\alpha + q'(x). \tag{IV.11.7}$$

As in the previous section we can find a relation for $\hat{u}_2(x, x)$ by differentiating in (IV.11.1–IV.11.4) with respect to x. Upon doing this, we obtain the system

$$\hat{u}_{21}(z, x) = a(z)\hat{u}_2(z, x) + b(z)\hat{v}_2(z, x), \tag{IV.11.8}$$

$$-\hat{v}_{21}(z, x) = c(z)\hat{u}_2(z, x) + d(z)\hat{v}_2(z, x), \tag{IV.11.9}$$

subject to

$$\hat{u}_2(0, x) = 0, \tag{IV.11.10}$$

$$\hat{v}_2(x, x) = -\hat{v}_1(x, x). \tag{IV.11.11}$$

Notice that the functions $\hat{u}_2(z, x)$ and $\hat{v}_2(z, x)$ and the functions $u(z, x)$ and $v(z, x)$ of the previous section satisfy the same differential equations, but with slightly different boundary conditions.

According to exercise 13 of Chapter III, the relationship between $u(z, x)$ and $\hat{u}_2(z, x)$ is given by

$$\hat{u}_2(z, x) = -\hat{v}_1(x, x)u(z, x). \tag{IV.11.12}$$

Using (IV.11.2), evaluated at $z=x$, and then (IV.11.4) and (IV.11.5), then (IV.11.12) becomes

$$\hat{u}_2(z, x) = \alpha[c(x)r(x) + d(x)]u(z, x) + [c(x)q(x) + f(x)]u(z, x). \tag{IV.11.13}$$

If we evaluate (IV.11.13) at $z=x$ and use (IV.11.5) and substitute into (IV.11.7) and finally equate coefficients of "α" and "1" we find that

$$r'(x) = b(x) + [a(x) + d(x)]r(x) + c(x)r^2(x), \qquad \text{(IV.11.14)}$$

$$q'(x) = [a(x) + c(x)r(x)]q(x) + f(x)r(x) + e(x), \qquad \text{(IV.11.15)}$$

subject to

$$r(0) = 0, \qquad \text{(IV.11.16)}$$

$$q(0) = 0. \qquad \text{(IV.11.17)}$$

In order to obtain the internal values, we define

$$\hat{u}(z, x) = J(z, x)\alpha + p(z, x). \qquad \text{(IV.11.18)}$$

Then, by differentiating in (IV.11.18) with respect to x, we get

$$\hat{u}_2(z, x) = J_2(z, x)\alpha + p_2(z, x). \qquad \text{(IV.11.19)}$$

Again comparing the coefficients of "α" and "1" in (IV.11.13) and (IV.11.19), we obtain the differential equations

$$J_2(z, x) = [c(x)r(x) + d(x)]J(z, x), \qquad \text{(IV.11.20)}$$

$$p_2(z, x) = [c(x)q(x) + f(x)]J(z, x), \qquad \text{(IV.11.21)}$$

subject to

$$J(z, z) = r(z), \qquad \text{(IV.11.22)}$$

$$p(z, z) = q(z). \qquad \text{(IV.11.23)}$$

The algorithm is very similar to that described in the previous section, except that we now must integrate two equations initially (IV.11.14) and (IV.11.15) and, at each desired internal point, two additional equations (of the form (IV.11.20) and (IV.11.21)) must be adjoined to the system. The internal values are then obtained by the use of (IV.11.18).

12   The Method of Kagiwada and Kalaba – Troubles

In the two previous sections we dealt with a special type of boundary conditions; namely,

$$u(0, x) = 0, \quad v(x, x) = \alpha. \qquad \text{(IV.12.1)}$$

Although many problems are formulated with these boundary conditions or can be transformed into this form (see exercise 11, Chapter III), a very important class of problems has boundary conditions of the form

$$u\,(0,\,x) = 0, \quad u\,(x,\,x) = 1. \tag{IV.12.2}$$

Boundary conditions of this type are most commonly associated with a second-order equation of the form

$$u''\,(z) + m\,(z)\,u'\,(z) + n\,(z)\,u\,(z) = 0. \tag{IV.12.3}$$

We shall, however, continue to consider the two first-order equations

$$u_1\,(z,\,x) = a\,(z)\,u\,(z,\,x) + b\,(z)\,v\,(z,\,x), \tag{IV.12.4}$$

$$-\,v_1\,(z,\,x) = c\,(z)\,u\,(z,\,x) + d\,(z)\,v\,(z,\,x), \tag{IV.12.5}$$

subject to

$$u\,(0,\,x) = 0, \tag{IV.12.6}$$

$$u\,(x,\,x) = 1. \tag{IV.12.7}$$

Since $u(x,\,x)$ is given for this problem, we must modify our previous analysis. We thus define

$$s\,(x) = v\,(z,\,x)\big|_{z=x} = v\,(x,\,x). \tag{IV.12.8}$$

Proceeding as before, we find that $s(x)$ satisfies the differential equation

$$-\,s'\,(x) = c\,(x) + [a\,(x) + d\,(x)]\,s\,(x) + b\,(x)\,s^2\,(x). \tag{IV.12.9}$$

The problem is not in the derivation of the above differential equation, but in the initial condition for $s(0)$. Since $u(0,\,x)=0$, and $u(x,\,x)=1$, it is clear that

$$\frac{du\,(z,\,x)}{dz} \to \infty \quad \text{as} \quad x \to 0. \tag{IV.12.10}$$

It then follows from (IV.12.4) that

$$b\,(z)\,v\,(z,\,x) \to \infty \quad \text{as} \quad x \to 0. \tag{IV.12.11}$$

Consequently, the initial condition for s is

$$s\,(0) = \begin{cases} +\infty, & b\,(0) > 0. \\ -\infty, & b\,(0) < 0. \end{cases} \tag{IV.12.12}$$

By letting $w(x) = v(0, x)$, we find that $w(x)$ satisfies

$$w'(x) = -[a(x) + b(x)s(x)]w(x) \qquad \text{(IV.12.13)}$$

with

$$w(0) = \begin{cases} +\infty, & b(0) > 0 \\ -\infty, & b(0) < 0. \end{cases} \qquad \text{(IV.12.14)}$$

Analytically the infinite initial conditions cause no particular problem. However, from a numerical standpoint the infinite initial conditions are not very desirable. We have to replace infinity by some large number (usually the largest number accepted by the computer) and hope that the solution for large x is not sensitive to the value chosen for the initial conditions. It happens that many of the Riccati equations of invariant imbedding possess the property. We shall deal with this in more detail later but, for now, let us consider the particular example

$$u_1(z, x) = v(z, x), \qquad \text{(IV.12.15)}$$

$$-v_1(z, x) = -u(z, x) \qquad \text{(IV.12.16)}$$

with

$$u(0, x) = 0, \qquad \text{(IV.12.17)}$$

$$u(x, x) = 1. \qquad \text{(IV.12.18)}$$

For this problem (IV.12.9) becomes

$$-s'(x) = -1 + s^2(x) \qquad \text{(IV.12.19)}$$

and, since $b(0) \equiv 1$, the initial condition is

$$s(0) = +\infty. \qquad \text{(IV.12.20)}$$

Numerically, we must, of course, replace (IV.12.19, IV.12.20) by

$$-s_k'(x) = -1 + s_k^2(x), \qquad \text{(IV.12.21)}$$

subject to

$$s_k(0) = K, \qquad \text{(IV.12.22)}$$

where K is a large positive number. The solution of (IV.12.21, IV.12.22) is given by

$$s_k(x) = \operatorname{cotanh}(x + \operatorname{cotanh}^{-1} K). \qquad \text{(IV.12.23)}$$

In Table IV.4-1 we tabulated the solution of (IV.12.19–IV.12.22), for $K=10$, 100, and ∞ to illustrate how rapidly the solution with a large initial value approaches the solution with infinity as an initial condition.

TABLE IV.4-1.   *Solution of $s'=1-s^2$, $s(0)= K$*
for Various Values of K

x	$S_{10}(x)$	$S_{100}(x)$	$S_k(x)$
0.0	10	100	∞
0.1	3.34	9.14	10.05
0.2	2.58	4.83	5.08
0.5	1.39	2.14	2.17
0.75	1.35	1.56	1.58
1.00	1.20	1.305	1.31
1.50	1.07	1.103	1.105
2.00	1.03	1.040	1.040
3.00	1.003	1.005	1.005

13 Numerical Examples

In this section we shall discuss some of the practical aspects of the numerical solution of the invariant imbedding equations. Several pertinent examples will be presented.

In any actual implementation of the versions of invariant imbedding developed in this chapter, there will be several programming decisions. For example, the generalized version described in section 5 requires the integration of more differential equations than does the basic algorithm of section 2. However, if a parametric study for various boundary conditions is desired or if the boundary conditions are given by (IV.6.1, IV.6.2), the generalized version may be more appropriate. Similarly, one can either obtain the values of $u(z)$ and $v(z)$ by the basic algorithm (or the generalized algorithm) or use the method discussed in exercise 1. The first alternative would minimize the use of computational time and the second would economize on computer storage. Decisions on these matters should be made on the basis of the type of problem to be solved, the type and amount of information desired, and the characteristics of the particular computer system available.

One of the primary sources of error in the method of invariant imbedding is, curiously enough, not in numerical integration of the differential equations but in the additions performed in (IV.2.19) and (IV.2.20). This problem can be easily illustrated by considering the analytic example of section 4. We have the problem

$$y'' = y + 1, \tag{IV.13.1}$$

$$y(0) = \alpha, \quad y'(x) = \beta. \tag{IV.13.2}$$

From (IV.2.19) one of the basic quantities desired is the term $[q_2(x) - q_2(z)]$. The quantity $q_2(z)$ for the above problem is given by (IV.4.14) as

$$q_2(z) = -(\alpha + 1) \tanh z. \tag{IV.13.3}$$

For any value of $x > 10$, a large number of accurate digits in the numerical integration of $q_2(z)$ would be required. For example, let $x = 20$ and $z = 18$. Then,

$$\begin{aligned} q_2(x) - q_2(z) &= -(\alpha + 1)(\tanh x - \tanh z) \\ &= -(\alpha + 1)(4.554 \times 10^{-16}). \end{aligned} \tag{IV.13.4}$$

Thus, even if the exact solution is known, the numerical evaluation can be erroneous if a sufficient number of significant digits is not retained by the com-

TABLE IV.4-2.   *Exact Solution of* $y'' = y + 1$, $y(0) = 0$, $y'(20) = 1$ *Using Single-Precision Arithmetic on EMR-6130*

z	$q_2(20) - q_2(z)$	$v = y'$	$u = y$
0.0	0.1000000E 01	−0.1000000E 01	0.0000000E 00
1.0	0.2384057E 00	−0.3678791E 00	−0.6321201E 00
2.0	0.3597259E−01	−0.1353360E 00	−0.8646655E 00
3.0	0.4945278E−02	−0.4978734E−01	−0.9502134E 00
4.0	0.6709099E−03	−0.1832126E−01	−0.9816904E−00
5.0	0.9059906E−04	−0.6723046E−02	−0.9932475E 00
6.0	0.1239777E−04	−0.2499992E−02	−0.9975429E 00
7.0	0.1430511E−05	−0.7821132E−03	−0.9989586E 00
8.0	0.0000000E 00	0.6144211E−05	−0.9993234E 00
9.0	0.0000000E 00	0.1670170E−04	−0.9997368E 00
10.0	0.0000000E 00	0.4539992E−04	−0.9998641E 00
11.0	0.0000000E 00	0.1234098E−03	−0.9998436E 00
12.0	0.0000000E 00	0.3354625E−03	−0.9996526E 00
13.0	0.0000000E 00	0.9118817E−03	−0.9990840E 00
14.0	0.0000000E 00	0.2478752E−02	−0.9975200E 00
15.0	0.0000000E 00	0.6737947E−02	−0.9932618E 00
16.0	0.0000000E 00	0.1831564E−01	−0.9816844E 00
17.0	0.0000000E 00	0.4978706E−01	−0.9502132E 00
18.0	0.0000000E 00	0.1353353E 00	−0.8646650E 00
19.0	0.0000000E 00	0.3678793E 00	−0.6321211E 00
20.0	0.0000000E 00	0.1000000E 01	0.0000000E 00

TABLE IV.4-3.   *Exact Solution of* $y''=y+1$, $y(0)=0$, $y'(20)=1$ *Using Double-Precision Arithmetic on EMR-6130*

z	$q_2(20)-q_2(z)$		$v=y'$		$u=y$	
0.0	0.1000000D	01	−0.1000000D	01	0.5684342D−13	
1.0	0.2384058D	00	−0.3678794D	00	−0.6321206D	00
2.0	0.3597242D−01		−0.1353353D	00	−0.8646647D	00
3.0	0.4945246D−02		−0.4978703D−01		−0.9502129D	00
4.0	0.6707003D−03		−0.1831553D−01		−0.9816842D	00
5.0	0.9079574D−04		−0.6737641D−02		−0.9932617D	00
6.0	0.1228835D−04		−0.2477921D−02		−0.9975204D	00
7.0	0.1663056D−05		−0.9096216D−03		−0.9990859D	00
8.0	0.2250702D−06		−0.3293182D−03		−0.9996584D	00
9.0	0.3045989D−07		−0.1067078D−03		−0.9998599D	00
10.0	0.4122285D−08		0.2478977D−09		−0.9999092D	00
11.0	0.5578613D−09		0.1067091D−03		−0.9998599D	00
12.0	0.7548806D−10		0.3293196D−03		−0.9996584D	00
13.0	0.1011813D−10		0.9096438D−03		−0.9990858D	00
14.0	0.1364242D−11		0.2477932D−02		−0.9975204D	00
15.0	0.1136868D−12		0.6737761D−02		−0.9932616D	00
16.0	0.0000000D	00	0.1831564D−01		−0.9816841D	00
17.0	0.0000000D	00	0.4978707D−01		−0.9502128D	00
18.0	0.0000000D	00	0.1353353D	00	−0.8646647D	00
19.0	0.0000000D	00	0.3678794D	00	−0.6321205D	00
20.0	0.0000000D	00	0.1000000D	01	0.4122342D−08	

puter. This is illustrated by Tables IV.4-2 and IV.4-3. In Table IV.4-2 we have evaluated the exact solution of the above problem for $x=20$ in single precision on an EMR-6130 computer, which carries about six significant digits. In Table IV.4-3 the same results are listed, using double precision. Notice that, even though $u=y$ is accurate in both tables, $v=y'$ is of opposite sign at $z=8$ and $z=9$.

In order to minimize the effect of the cancellation error associated with the calculation of $q_2(x)-q_2(z)$, we recommend the following procedure: select values $0=z_0<z_1<\cdots<z_n=x$. Replace (IV.2.15) by the initial-value problem

$$w_i' = [c(z)r_2(z) + f(z)]q_1(z), \qquad \text{(IV.13.5)}$$

$$w_i(z_{i-1}) = 0, \qquad \text{(IV.13.6)}$$

for z in (z_{i-1}, z_i). The value of $q_2(x)-q_2(z)$ at the desired values of z is then to be computed by the formula

$$q_2(x) - q_2(z) = [w_i(z_i) - w_i(z)]$$
$$+ w_{i+1}(z_{i+1}) + \cdots + w_n(z_n), \quad z_{i-1} < z \leqslant z_i. \quad \text{(IV.13.7)}$$

The adoption of this procedure involves little additional programming effort and no additional computational cost. Nelson has termed this procedure as the *method of successive starts.*

We shall discuss the actual numerical computation of (IV.13.1) and (IV.13.2) using several formulations of the problem and different computers. Let $\alpha = 0$ and $\beta = 1$ and $x = 20.0$. In all cases a classical fourth-order Runge-Kutta integration scheme with the fixed step-size of $\Delta z = 0.2$ was used. Although, in practice, it would

TABLE IV.4-4.   *Numerical Solution of Invariant Imbedding Equations on EMR-6130.*

z	$R_1(z)$	$R_2(z)$	$R_3(z)$
0.0	0.000000E 00	0.100000E 01	0.000000E 00
2.0	0.964007E 00	0.265802E 00	−0.734199E 00
4.0	0.999328E 00	0.366204E−01	−0.963380E 00
6.0	0.999987E 00	0.495782E−02	−0.995043E 00
8.0	0.100000E 01	0.670993E−03	−0.999330E 00
10.0	0.100000E 01	0.908118E−04	−0.999910E 00
12.0	0.100000E 01	0.122904E−04	−0.999989E 00
14.0	0.100000E 01	0.166338E−05	−0.100000E 01
16.0	0.100000E 01	0.225121E−06	−0.100000E 01
18.0	0.100000E 01	0.304678E−07	−0.100000E 01
20.0	0.100000E 01	0.412350E−08	−0.100000E 01

z	$Q_1(z)$	$Q_2(z)$	$Q_3(z)$
0.0	0.100000E 01	0.000000E 00	0.000000E 00
2.0	0.265802E 00	−0.964046E 00	−0.964047E 00
4.0	0.366204E−01	−0.999351E 00	−0.999351E 00
6.0	0.495782E−02	−0.100001E 01	−0.100001E 01
8.0	0.670993E−03	−0.100002E 01	−0.100002E 01
10.0	0.908118E−04	−0.100002E 01	−0.100002E 01
12.0	0.122904E−04	−0.100002E 01	−0.100002E 01
14.0	0.166338E−05	−0.100002E 01	−0.100002E 01
16.0	0.225121E−06	−0.100002E 01	−0.100002E 01
18.0	0.304678E−07	−0.100002E 01	−0.100002E 01
20.0	0.412350E−08	−0.100002E 01	−0.100002E 01

be desirable to use a variable step integration scheme, our basic intent here is to show the effect of various formulations and computer upon the results of certain problems.

The first computer used was an EMR-6130, which is capable of yielding approximately six significant digits. The second computer was a CDC-6600 with a 60 bit word length, or approximately thirteen significant digits. The computed results for the r and q functions are shown in Table IV.4-4 for the EMR-6130 and in Table IV.4-5 for the CDC-6600. (Since $u(0)=\alpha=0$, it is not necessary to compute $r_2(z)$ and $q_2(z)$; however, they are shown here for illustration.) These results show very little difference and, in fact, both are accurate to about five significant figures. The differential equations for the r and q functions are quite stable and easily integrated. However, as expected from the discussion above, the real error is introduced in the computation of $u(z)$ and $v(z)$ from

TABLE IV.4-5.   *Numerical Solution of Invariant Imbedding Equations on CDC-6600*

z	$R_1(z)$	$R_2(z)$	$R_3(z)$
0.0	0.0	0.100000E+01	0.0
2.0	0.964008E−00	0.265802E−00	−0.734198E−00
4.0	0.999328E−00	0.366204E−01	−0.963379E−00
6.0	0.999988E−00	0.495782E−02	−0.995042E−00
8.0	0.900000E+00	0.670993E−03	−0.999329E−00
10.0	0.100000E+01	0.908119E−04	−0.999909E−00
12.0	0.100000E+01	0.122904E−04	−0.999988E−00
14.0	0.100000E+01	0.166338E−05	−0.999998E−00
16.0	0.100000E+01	0.225122E−06	−0.100000E+01
18.0	0.100000E+01	0.304679E−08	−0.100000E+01
20.0	0.100000E+01	0.412351E−08	−0.100000E+01

z	$Q_1(z)$	$Q_2(z)$	$Q_3(z)$
0.0	0.100000E+00	0.0	0.0
2.0	0.265802E−00	−0.964047E−00	−0.964047E−00
4.0	0.366204E−01	−0.999350E−00	−0.999350E−00
6.0	0.495782E−02	−0.100000E+01	−0.100001E+01
8.0	0.670993E−03	−0.100002E+01	−0.100002E+01
10.0	0.908119E−04	−0.100002E+01	−0.100002E+01
12.0	0.122905E−04	−0.100002E+01	−0.100002E+01
14.0	0.166338E−05	−0.100002E+01	−0.100002E+01
16.0	0.225122E−07	−0.100002E+01	−0.100002E+01
18.0	0.304679E−07	−0.100002E+01	−0.100002E+01
20.0	0.412351E−08	−0.100002E+01	−0.100002E+01

(IV.2.19) and (IV.2.20). The EMR and the CDC results are shown, respectively, in Tables IV.4-6 and IV.4-7. Comparing these results with those given in Table IV.4-3 we see that the EMR results are meaningless for $z>12$; whereas the CDC results, due to the longer word length, are good for all z.

In Table IV.4-8 we have displayed the results of the method of successive starts as discussed above. Notice that there is a substantial improvement in the accuracy of the results. It should be re-emphasized that there is no extra computing time involved in the use of the method of successive starts and in no case will it introduce error.

In terms of accuracy and stability the algorithm by Kagiwada and Kalaba, as introduced in sections 10-12, is one of the very best available for solving linear two-point boundary-value problems. Its primary drawback is the large number of integrations involved if many internal points are desired. This can become quite time consuming, especially for systems of equations. The results, using the EMR, are shown in Table IV.4-9 and are accurate to about five significant figures.

TABLE IV.4-6.   *Invariant Imbedding Solution of $y''=y+1$, $y(0)=0$, $y'(20)=1$ Using Single-Precision Arithmetic on EMR-6130*

z	u(z)	v(z)	w(z)
0.0	0.412341E−08	−0.100002E 01	−0.100002E 01
2.0	−0.864676E 00	−0.135349E 00	−0.100002E 01
4.0	−0.981702E 00	−0.183337E−01	−0.100004E 01
6.0	−0.997447E 00	−0.240447E−02	−0.999852E 00
8.0	−0.998619E 00	0.710644E−03	−0.997909E 00
10.0	−0.994659E 00	0.525083E−02	−0.989409E 00
12.0	−0.961191E 00	0.387974E−01	−0.922394E 00
14.0	0.713332E 00	0.286667E 00	−0.426665E 00
16.0	0.111813E 01	0.211813E 01	0.323627E 01
18.0	0.146505E 02	0.156505E 02	0.303010E 02
20.0	0.114639E 03	0.115639E 03	0.230278E 03

TABLE IV.4-7.   *Invariant Imbedding Solution of $y''=y+1$, $y(0)=0$, $y'(20)=1$ Using Single-Precision Arithmetic on CDC-6600*

z	u(z)	v(z)	w(z)
0.0	−0.135994E−14	−0.100002E+01	−0.100002E+01
2.0	−0.864668E−00	−0.135341E−00	−0.100001E+01
4.0	−0.986184E−00	−0.183169E−01	−0.100000E+01
6.0	−0.997520E−00	−0.247818E−02	−0.999998E−00
8.0	−0.999658E−00	−0.329363E−03	−0.999987E−00
10.0	−0.999909E−00	0.101716E−08	−0.999909E−00
12.0	−0.999658E−00	0.329368E−03	−0.999329E−00
14.0	−0.997520E−00	0.247819E−02	−0.995041E−00
16.0	−0.981683E−00	0.183168E−01	−0.963366E−00
18.0	−0.864650E−00	0.135339E−00	−0.729320E−00
20.0	0.218025E−05	0.100000E+01	0.100000E+01

TABLE IV.4-8.   *Invariant Imbedding Solution of $y''=y+1$, $y(0)=0$, $y'(20)=1$ Using the Method of Successive Starts and Single-Precision Arithmetic on EMR-6130*

z	u(z)	v(z)	w(z)
0.0	0.000000E 00	−0.100002E 01	−0.100002E 01
2.0	−0.864668E 00	−0.135341E 00	−0.100001E 01
4.0	−0.981684E 00	−0.183164E−01	−0.100000E 01
6.0	−0.997521E 00	−0.247752E−02	−0.999998E 00
8.0	−0.999659E 00	−0.328593E−03	−0.999998E 00
10.0	−0.999909E 00	0.147536E−05	−0.999907E 00
12.0	−0.999653E 00	0.335505E−03	−0.999318E 00
14.0	−0.997481E 00	0.251873E−02	−0.994962E 00
16.0	−0.981385E 00	0.186150E−01	−0.962770E 00
18.0	−0.862452E 00	0.137548E 00	−0.724905E 00
20.0	0.163193E−01	0.101632E 01	0.103264E 01

TABLE IV.4-9.   *Invariant Imbedding Solution of $y''=y+1$, $y(0)=0$, $y'(20)=1$ Using the Method Kagiwada and Kalaba and Single-Precision Arithmetic on EMR-6130*

z	u(z, x)
0.0	0.000000E 00
1.0	−0.632132E 00
2.0	−0.864671E 00
3.0	−0.950217E 00
4.0	−0.981687E 00
5.0	−0.993264E 00
6.0	−0.997522E 00
7.0	−0.999088E 00
8.0	−0.999660E 00
9.0	−0.999861E 00
10.0	−0.999910E 00
11.0	−0.999861E 00
12.0	−0.999659E 00
13.0	−0.999086E 00
14.0	−0.997521E 00
15.0	−0.993261E 00
16.0	−0.981683E 00
17.0	−0.950210E 00
18.0	−0.864660E 00
19.0	−0.632115E 00
20.0	0.238419E−06

As we have already stated, the primary source of error in the method of invariant imbedding, as discussed earlier in this section, is in the computation of $q_2(x) - q_2(z)$. If the source terms are zero and if $u(0) = 0$, then $q_2(z) = 0$. Hence, there is no contribution of error from this term. In this case, the solutions for $u(z)$ and $v(z)$ are quite accurate. This is illustrated by considering the equation

$$y'' = y, \tag{IV.13.8}$$

subject to

$$y(0) = 0, \quad y'(x) = 1. \tag{IV.13.9}$$

The computed results and the analytical results are listed, respectively, in

TABLE IV.4-10.   *Invariant Imbedding Solution of $y'' = y$, $y(0) = 0$, $y'(20) = 1$ Using Single-Precision Arithmetic on EMR-6130*

z	$u(z)$	$v(z)$	$w(z)$
0.0	0.177636E−14	0.412350E−08	0.412350E−08
2.0	0.149551E−07	0.155134E−07	0.304685E−07
4.0	0.112526E−06	0.112601E−06	0.225127E−06
6.0	0.831707E−06	0.831718E−06	0.166342E−05
8.0	0.614538E−05	0.614538E−05	0.122908E−04
10.0	0.454071E−04	0.454071E−04	0.908143E−04
12.0	0.335505E−03	0.335505E−03	0.671010E−03
14.0	0.247899E−02	0.247899E−02	0.495798E−02
16.0	0.183168E−01	0.183168E−01	0.366336E−01
18.0	0.135339E 00	0.135340E 00	0.270679E 00
20.0	0.100000E 01	0.100000E 01	0.200000E 01

TABLE IV.4-11   *Exact Solution of $y'' = y$, $y(0) = 0$, $y'(20) = 1$*

z	$u(z)$	$v(z)$	$w(z)$
0.0	0.000000E 00	0.412231E−08	0.412231E−08
2.0	0.149510E−07	0.155089E−07	0.304600E−07
4.0	0.112497E−06	0.112573E−06	0.225070E−06
6.0	0.831523E−06	0.831534E−06	0.166306E−05
8.0	0.614421E−05	0.614421E−05	0.122884E−04
10.0	0.453999E−04	0.453999E−04	0.907998E−04
12.0	0.335463E−03	0.335463E−03	0.670925E−03
14.0	0.247875E−02	0.247875E−02	0.495750E−02
16.0	0.183156E−01	0.183156E−01	0.366313E−01
18.0	0.135335E 00	0.135335E 00	0.270671E 00
20.0	0.100000E 01	0.100000E 01	0.200000E 01

Tables IV.4-10 and IV.4-11. Both results were obtained using single precision arithmetic on an EMR-6130 computer.

We shall now present two numerical examples to illustrate the use of the procedures of section 8 in calculating Green's functions. The examples were run on an EMR-6130 computer. The integration procedure used was a simple fourth-order Runge-Kutta routine with a fixed step size of $h=0.01$. Obviously, our results are of a feasibility nature and not designed to obtain the ultimate in accuracy.

In our first example we wish to calculate the Green's function for the operator L, where

$$Ly = y'' - y = 0, \tag{IV.13.10}$$

subject to the boundary conditions

$$y(0) = 0, \quad y'(x) = 0. \tag{IV.13.11}$$

This operator was chosen since it can be easily resolved in closed form for comparison. The analytical solution is given by

$$g(z, z', x) = \frac{\sinh z' \cosh(x - z)}{\cosh x}, \quad z' < z, \tag{IV.13.12}$$

where $g(z, z', x)$ is symmetric in z and z'.

The equations for r_1 and r_4 are easily seen to be

$$r_1'(z) = 1 + r_1^2(z), \quad r_1(0) = 0, \tag{IV.13.13}$$

$$r_4'(z, z') = r_1(z) r_4(z, z'), \quad r_4(z', z') = r_1(z'), \tag{IV.13.14}$$

whose solutions are

$$r_1(z) = \tanh z, \tag{IV.13.15}$$

$$r_4(z, z') = \sinh z' \operatorname{sech} z. \tag{IV.13.16}$$

The differential equation for $g(z, z', x)$ is given by

$$g'(z, z', x) = \cotanh z \, g(z, z', x) - \frac{\sinh z'}{\sinh z}, \tag{IV.13.17}$$

subject to the condition

$$g(x, z', x) = r_4(x, z'). \tag{IV.13.18}$$

The results for $x=1$ and a few arbitrarily selected values of z and z' are listed in Table IV.4–12. In Table IV4-12, $g(z, z', x)$ represents the numerical solution obtained by integrating (IV.13.17-IV.13.18), and $g_a(z\, z', x)$ is the analytical solution represented by (IV.13.12).

TABLE IV.4-12.   *Numerical Solution of the Green's Function for* $y'' = y$, $y(0)=0$, $y'(x) = 0$, *Using an Initial-Value Procedure*

z	z'	$g_c(z, z', x)$	$g_a(z, z', x)$
0.1	0.1	0.093024	0.093030
0.7	0.2	0.136389	0.136395
0.7	0.5	0.353013	0.353003
0.8	0.5	0.344470	0.344479
0.9	0.1	0.065240	0.065237
0.9	0.9	0.668561	0.668567

The second example is

$$y'' + y = 0,$$
(IV.13.19)

subject to

$$y(0) = 0, \qquad y'(x) = 0$$
(IV.13.20)

Again, the problem can be easily resolved analytically. In fact, we need only replace the hyberbolic functions in the first example by regular trigonometric functions. That is,

$$g(z, z', x) = \frac{\sin z' \cos(x - z)}{\cos x}, \qquad z' < z$$
(IV.13.21)

TABLE IV.4-13.   *Numerical Solution of the Green's Function for* $y' = -y$, $y(0) = 0$, $y'(x) = 0$ *Using an Initial-Value Procedure*

z	z'	$g_c(z, z', x)$	$g_a(z, z', x)$
0.1	0.1	0.114852	0.114853
0.7	0.2	0.351368	0.351281
0.7	0.5	0.847676	0.847712
0.8	0.5	0.869621	0.869656
0.9	0.1	0.183846	0.183844
0.9	0.9	1.44252	1.44256

$$r_1(z) = \tan z \qquad\qquad \text{(IV.13.22)}$$

$$r_4(z) = \sin z' \sec z \qquad\qquad \text{(IV.13.23)}$$

and, again, the Green's function is symmetric in z and z'. The numerical results are compared with the analytical results in Table IV.4–13 for $x=1$.

14 Exercises

1. Show that the following procedure is a valid alternative to the development of the recovery transformation. Integrate equations (IV.2.8) and (IV.2.9) subject to (IV.2.11) and (IV.2.12) from $z=0$ to $z=x$. Evaluate (IV.2.5) at $z=x$ to calculate the "missing" unknown boundary condition $u(x, x)$. Then, using (IV.2.5), write (IV.2.2) as

$$-\frac{\partial v(z, x)}{\partial z} = [c(z) r_1(z) + d(z)] v(z, x) + c(z) r_2(z) + f(z), \qquad (1)$$

subject to $v(x, x)=\beta$. Integrate this equation in the backward direction from $z=x$ to $z=0$. Use (IV.2.5) to calculate $u(z, x)$. Notice that every time the value of x is changed, (1) must be integrated backwards again. The method described here is frequently referred to as the *method of sweeps*. See I. Babuš-ka, M. Práger, and E. Vitásek, *Numerical Processes in Differential Equations*, Interscience Publishers, New York, 1966.

2. Derive (IV.2.14–IV.2.17).

3. Use the basic algorithm to solve the problem

$$y'' = y + 1 \qquad 0 \leqslant z \leqslant x \leqslant X$$
$$y(0) = 0, \qquad y'(x) = 1.$$

What restrictions, if any, must be put on X? (Hint: First convert to a system of two first order equations by letting $u=y$ and $v=y'$.)

4. Use the basic algorithm to solve the problem

$$y'' = -y + 1$$
$$y(0) = 0, \qquad y'(x) = 1, \qquad 0 \leqslant z \leqslant x \leqslant X.$$

What restrictions, if any, must be put on X?

5. Derive equations (IV.4.7–IV.4.13).

6. Derive equations (IV.5.7–IV.5.12).

7. Using the techniques of section 10 let

$$t(x) = v(0, x)$$

and derive the differential equation and initial condition satisfied by $t(x)$.

8. Derive the equations (IV.8.13, IV.8.14).

9. Verify the formula given in (IV.13.7).

15   Bibliographical Discussion

Section 2

A slightly modified version of this algorithm was first introduced in

M. R. Scott, "Invariant Imbedding and the Calculation of Internal Values," *J. Math. Anal. Appl.* **28** (1969), 112–119.

Related techniques not discussed in this book may be found in

E. D. Denman, *Coupled Modes in Plasmas, Electric Media, and Parametric Amplifiers*, American Elsevier Pub. Co., New York, 1970.
R. C. Allen, Jr. and G. M. Wing, "A Numerical Algorithm Suggested by Problems of Transsport in Periodic Media," *J. Math. Anal. Appl.* **29** (1970), 141–157.

Section 5

The generalized algorithm first appeared in

M. R. Scott, "Numerical Solution of Unstable Initial Value Problems," *Computer Journal* **13** (1970), 397–400.

A discussion of the generalized algorithm to problems of particle transport can be found in

P. Nelson, Jr. and M. R. Scott, "Internal Values in Particle Transport by the Method of Invariant Imbedding," *J. Math. Anal. Appl.* **34** (1971), 628–643.

Section 7

The inverse transformations play a fundamental role in eigenvalue problems and in solving nonhomogeneous problems on interval lengths greater than the first characteristic length. See Chapters V, VI, VIII

M. R. Scott, L. F. Shampine, and G. M. Wing, "Invariant Imbedding and the Calculation of Eigenvalues for Sturm-Liouville Systems," *Computing* **4** (1969), 10–23.

Section 8

The idea of generalizing the transformations to the integral forms of (IV.8.3) and (IV.8.4) grew out of the work reported in

C. W. Maynard and M. R. Scott, "Invariant Imbedding of Linear Partial Differential Equations Via Generalized Riccati Transformations," *J. Math. Anal. Appl.* **36** (1971), 432–459.

C. W. Maynard and M. R. Scott, "Some Relationships Between Green's Functions and Invariant Imbedding," to appear in *J. Opt. Th. Appl.*

For a different form of the imbedding see

R. Huss, H. H. Kagiwada, and R. E. Kalaba, "A Cauchy System for the Green's Function and the Solution of a Two-Point Boundary Value Problem," *J. Frank Inst.* **291** (1971), 159–167.

H. H. Kagiwada and R. E. Kalaba, "A Practical Method for Determining Green's Functions Using Hadamard's Variational Formula," *J. Opt. Th. Appl.* **1** (1967), 33–39.

Section 9

For some numerical results related to neutron transport in a rod, see

P. Nelson, Jr. and M. R. Scott, "Internal Values in Particle Transport by the Method of Invariant Imbedding," *J. Math. Anal. Appl.* **34** (1971), 628–643.

Sections 10–12

See

R. E. Bellman, H. H. Kagiwada, and R. E. Kalaba, "Invariant Imbedding and the Numerical Integration of Boundary-Value Problems for Unstable Linear Systems of Ordinary Differential Equations," *Comm. A.C.M.* **10** (1967), 100–102.

H. H. Kagiwada and R. E. Kalaba, "A New Initial-Value Method for Internal Intensities in Radiative Transfer," *The Astrophys. J.* **147** (1967), 301–309.

Section 13

The method of successive starts was first described in

P. Nelson, Jr. and C. A. Giles, "A Useful Device for Certain Boundary Value Problems," *J. Comp. Phys.* **10** (1972), 374–378.

An interesting comparison of the computational effort and accuracy between the method of invariant imbedding and the method of superposition appears in

P. Nelson, Jr., "A Comparative Study of Invariant Imbedding and Superposition," *Intern. J. Computer Math.*, Section B, **3** (1972).

For further numerical examples concerning the numerical construction of Green's functions, see the references listed under section 8.

V

LINEAR BOUNDARY-VALUE
PROBLEMS–HOMOGENEOUS

1 Introduction

In many applications of mathematical physics we are faced with the problem of solving for the values of a parameter, λ, so that the equation

$$Ly(z) = -\lambda y(z), \tag{V.1.1a}$$

subject to the operational relations

$$Ky = 0, \quad \text{at the boundaries } z = a \text{ and } z = b, \tag{V.1.1b}$$

has nontrivial solutions. Here, L is the nth-order operator given by

$$Ly = a_0 y^{(n)} + a_1 y^{(n-1)} + \cdots + a_n y, \tag{V.1.2}$$

where, unless otherwise stated, the a_j are restricted to be real-valued functions of the real variable z and are "sufficiently differentiable" on the closed interval $a \leqslant z \leqslant b$ and $a_0(z) \neq 0$ on $[a, b]$. Here Ky denotes the relation $K_j y = 0, j = 1, 2, ...,$ n, where

$$K_j y = \sum_{k=1}^{n} \left(M_{jk} y^{(k-1)}(a) + N_{jk} y^{(k-1)}(b) \right), \quad j = 1, 2, ..., n, \tag{V.1.3}$$

and the M_{jk} and N_{jk} are constants. The problem (V.1.1a, b) is called an *eigenvalue problem*.

The trivial solution, that is, $y(z) \equiv 0$, is always a solution of (V.1.1a, b). If the parameter λ is so that a nontrivial solution exists, then λ is called an *eigenvalue* or *characteristic value* and the nontrivial solution is said to be an *eigenfunction* or *characteristic function*.

Another type of problem encountered frequently in mathematical physics can be described as follows: Find the values of x, the interval length, so that the equation

$$Ly = -\lambda y, \quad \lambda \text{ fixed}, \quad \text{(V.1.4a)}$$

subject to

$$Ky = 0 \quad \text{at} \quad z = 0 \text{ and } z = x, \quad \text{(V.1.4b)}$$

has a nontrivial solution. We shall call such a problem an *eigenlength* or *characteristic length problem*. The values of x for which (V.1.4a, b) has a nontrivial solution are called *eigenlengths* or *characteristic lengths* and, again, the nontrivial solutions are referred to as *eigenfunctions* or *characteristic functions*.

Since our primary purpose will be to develop a mathematical technique which is amenable to numerical techniques for solving both the eigenvalue and eigenlength problems, we shall make the following assumptions:

(1) *The Eigenvalue Problem* – All the functions $a_j(z)$ and the constants M_{jk} and N_{jk} are such that the eigenvalues $\{\lambda_i\}$ are all real and form a denumerable set with no finite point of accumulation.
(2) *The Eigenlength Problem* – For fixed real λ, there are, at most, a denumerable number of interval lengths, $\{x_i\}$, so that a nontrivial solution exists.

Certainly these assumptions are not too restrictive since, if our problems are self-adjoint, the above do hold. Since the functions $a_j(z)$ are assumed to be real, the operator L is self-adjoint if, and only if, n is even. Hence, we shall be primarily concerned with even-ordered equations. Also, there are certain restrictions on the boundary conditions for the *problem* to be self-adjoint.

Although both of the above problems were formulated in terms of a single nth-order equation, we shall be primarily interested in a system formulation. That is, we shall consider systems of the form

$$u'(z) = A(z, \lambda) u(z) + B(z, \lambda) v(z), \quad \text{(V.1.5a)}$$

$$-v'(z) = C(z, \lambda) u(z) + D(z, \lambda) v(z), \quad \text{(V.1.5b)}$$

subject to

$$\alpha_1(\lambda) u(a) + \beta_1(\lambda) v(a) + \gamma_1(\lambda) u(b) + \delta_1(\lambda) v(b) = 0, \quad \text{(V.1.6a)}$$

$$\alpha_2(\lambda) u(a) + \beta_2(\lambda) v(a) + \gamma_2(\lambda) u(b) + \delta_2(\lambda) v(b) = 0, \quad \text{(V.1.6b)}$$

where A, B, C, D are $m \times m$ matrix functions of z; λ, α_i, β_i, γ_i, δ_i ($i = 1, 2$) are

$m \times m$ matrix functions of λ, the quantities u and v are m-rowed vector functions and $m = n/2$. The form of (V.1.5, V.1.6) is chosen mostly for convenience and historical reasons. This will become evident as we proceed.

Although the above problems are essentially equivalent, a point discussed later, there is a great deal of difference in the computational effort required to solve the two problems. The origins of eigenvalue problems are well documented, but there are considerably fewer documented characteristic length problems because most of the classical techniques were developed for eigenvalue problems rather than for characteristic length problems. Hence, most investigators force their problem into the eigenvalue category, whereas the more natural formulation might be a characteristic length problem. In order to illustrate the formulation of characteristic length problems we shall discuss two examples.

Our first example will come from the field of *Strength of Materials*, which is rich as a source of both types of problems. We wish to find the smallest length, x_1, of a rod so that it buckles under its own weight.

Let q [lb in^{-1}] be the mass per unit length of the rod, I [in^4] be the moment of inertia of the rod cross section, and E [lb in^{-2}] be the modulus of elasticity of the rod material. Then the bending moment at the point z of the rod (see Figure 5-1) is given by

$$M(z) = q \int_z^x \{y(s) - y(z)\} \, ds, \qquad (V.1.7)$$

where $y(z)$ is the elastic curve of the bent rod.

For small deflections of the rod we have

$$EI \frac{d^2 y}{dz^2} = M(z). \qquad (V.1.8)$$

Hence, upon substitution of (V.1.7) into (V.1.8), we obtain the integrodifferential equation

$$EI \frac{d^2 y}{dz^2} = q \int_z^x \{y(s) - y(z)\} \, ds. \qquad (V.1.9)$$

At $z = 0$, we have the conditions

$$y(0) = y'(0) = 0, \qquad (V.1.10)$$

and, from (V.1.9) evaluated at $z = x$, it follows that

$$y''(x) = 0. \qquad (V.1.11)$$

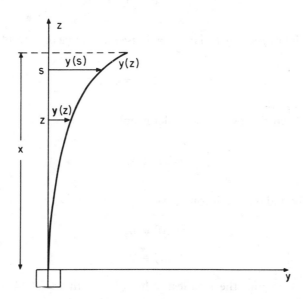

Figure 5-1.   The Bending of a Rod With Lower End Fixed and Upper End Free

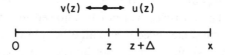

Figure 5-2.   One-Dimensional Rod Model for Neutron Transport

Differentiation in (V.1.9) with respect to z yields the third order equation

$$EI \frac{d^3 y}{dz^3} = -q \frac{dy}{dz} (x - z).$$

$$(V.1.12)$$

Let us now define $p = -q/EI$ [in^{-3}] and use the change of variables

$$w = \frac{dy}{dz} \quad \text{and} \quad t = x - z.$$

$$(V.1.13)$$

Then (V.1.12) becomes the second-order problem

$$\frac{d^2 w}{dt^2} = ptw(t),$$

$$(V.1.14)$$

subject to the boundary conditions

$$w'(0) = 0,$$

$$(V.1.15)$$

$$w(x) = 0.$$

$$(V.1.16)$$

Hence, we wish to find the smallest value of x so that (V.1.14–V.1.16) has a nontrivial solution, that is, we have a characteristic length problem.

Our second example, discussed in Chapter III and repeated here, deals with the time independent transport of particles, which we can think of as neutrons, in a one-dimensional rod. A simple model will be chosen to illustrate the idea and not necessarily to approximate a true physical situation. The neutrons will be allowed to traverse the rod in either direction, with a certain probability of interacting with the fixed nuclei of the rod. When an interaction does occur, the original neutron disappears and two others appear, one traveling to the right and one traveling to the left. The particles will not be allowed to interact among themselves, all travel with the same speed c, and all other physical properties are such that the particles are distinguishable only by their direction. The rod will be situated with its left end at $z=0$ and its right end at $z=x$.

We define

$u(z)=$ expected density of particles at z and moving to the right,
$v(z)=$ expected density of particles at z and moving to the left,
$\sigma\Delta +0(\Delta)=$ probability of a collision occurring between the rod and a particle moving between z and $z+\Delta$ in either direction.

The quantity σ is sometimes called the macroscopic cross section, and we shall assume it to be constant.

We shall now derive a set of differential equations satisfied by $u(z)$ and $v(z)$. Consider Figure 5-2.

The number of neutrons traveling to the right at z is a consequence of the number of neutrons at z which traverse the interval $[z, z+\Delta]$ without a collision, together with the consequence of interactions in $[z, z+\Delta]$ due to the terms $u(z)$ and $v(z+\Delta)$. Explicitly, we have

$$cu\,(z + \Delta) = cu\,(z) + cv\,(z + \Delta)\,\sigma\Delta + \mathrm{o}\,(\Delta),$$

where multiple interactions are included in the term $\mathrm{o}\,(\Delta)$. Assuming that $u(z)$ and $v(z)$ are continuous, we divide through by $c\Delta$ and let $\Delta \to 0$. We then obtain the differential equation

$$\frac{du}{dz} = \sigma v\,(z). \qquad\qquad \text{(V.1.17)}$$

Similarly for the left-moving particles, we find

$$cv\,(z) = cv\,(z + \Delta) + \sigma\Delta cu\,(z) + \mathrm{o}\,(\Delta),$$

and so

$$-\frac{dv}{dz} = \sigma u\,(z). \qquad\qquad \text{(V.1.18)}$$

For boundary conditions let us suppose that each second a single right-moving particle is injected into the rod at $z=0$ and no particles enter at $z=x$. Any particle escaping at $z=0$ or $z=x$ is permanently lost from the system. Thus,

$$cu\,(0) = 1, \qquad\qquad \text{(V.1.19)}$$

$$cv\,(x) = 0. \qquad\qquad \text{(V.1.20)}$$

The solution of (V.1.17–V.1.20) is, of course, quite simple and is given by

$$cu\,(z) = \frac{\cos \sigma\,(x - z)}{\cos \sigma x}, \qquad\qquad \text{(V.1.21)}$$

$$cv\,(z) = \frac{\sin \sigma\,(x - z)}{\cos \sigma x}. \qquad\qquad \text{(V.1.22)}$$

Since $cu(z)$ and $cv(z)$ are expected numbers of particles, we must require them to be nonnegative on $0 \leqslant z \leqslant x$. This will certainly be true provided that $x < \pi/2\sigma$. When $x = \pi/2\sigma$, the denominators of (V.1.21) and (V.1.22) become zero; hence, there is no solution to the nonhomogeneous problem defined by (V.1.17–V.1.20). However, this implies that the homogeneous problem (V.1.19) replaced by $cu(0)=0$ does have a nontrivial solution. Such a value of x is termed the "critical" length or, in our terminology, the first characteristic length.

The classical techniques for finding the critical length would involve reformulating the problem as

$$\frac{d^2u}{dz^2} + \lambda\sigma^2 u = 0,$$ (V.1.23a)

$$v = \frac{1}{\sigma}\frac{du}{dz},$$ (V.1.23b)

$$u(0) = 0,$$ (V.1.23c)

$$u'(x) = 0.$$ (V.1.23d)

Then, ignoring the fact that we have a closed form solution in this simple case, pick values of x until one is found so that $\lambda = 1$ is an eigenvalue belonging to non-negative functions $u(z)$ and $v(z)$. Obviously, several values of x will normally be required before the desired value can be found.

In this chapter we shall derive methods based upon the method of invariant imbedding, which calculate the characteristic lengths directly, and by using a simple iteration procedure, various types of eigenvalue problems. Included among these will be problems where the eigenvalue parameter appears in a non-linear fashion, various types of singularities, and in very general boundary conditions including periodic conditions. In addition, the procedure generalizes to systems in a straightforward manner. Inasmuch as the differential equations involved in the procedure are initial-value problems and tend to be quite stable numerically, they avoid many of the instabilities normally associated with procedures for solving two-point boundary-value problems.

2   The Characteristic Lengths and Their Calculation

We consider the system

$$\frac{du}{dz} = a(z)u(z) + b(z)v(z)$$ (V.2.1a)

$$-\frac{dv}{dz} = c(z)u(z) + d(z)v(z)$$ (V.2.1b)

$$u(0) = 0,$$ (V.2.1c)

$$u(x) = 0, \quad 0 \leqslant z \leqslant x.$$ (V.2.1d)

The coefficients a, b, c, and d are assumed to be continuous real functions on $0 \leqslant z \leqslant x \leqslant X$. This restriction is assumed for convenience at this point, since it assures us that existence and uniqueness hold for initial-value problems for (V.2.1a, b) on $0 \leqslant z \leqslant x \leqslant X$. In a later section we shall discuss formulations where the coefficients may be singular.

In general, the only solution of (V.2.1a, b) satisfying the boundary conditions is the trivial one, $u \equiv v \equiv 0$. However, there may be certain x values, $0 < x_1 < < x_2 < \cdots < x_n \leqslant X$, for which (V.2.1) has a nontrivial solution. We call such values x_i, the *characteristic lengths* of the system.

If u and v are nontrivial solutions of (V.2.1) when $x = x_i$, then $v(0) \neq 0$. Otherwise, by the assumed uniqueness we should have $u \equiv v \equiv 0$. We introduce the function $r(z)$ by the relation

$$u(z) = r(z) v(z) \qquad\qquad \text{(V.2.2)}$$

(The relation (V.2.2) had its inception in the study of invariant imbedding and, for the particle transport rod model in the introduction, the function $r(z)$ gives the relation between the right-moving and the left-moving particles. However, in the context of this section, we may simply think of (V.2.2) as a generalization of the well-known Riccati transformation).

Assume $v(z) \neq 0$ in $0 \leqslant z < z_1$. Then $r(z)$ is well defined there. Substituting into (V.2.1a), followed by the use of (V.2.1b) and (V.2.2), yields

$$\{r'(z) - b(z) - [a(z) + d(z)] r(z) - c(z) r^2(z)\} v(z) = 0. \qquad \text{(V.2.3)}$$

Since $v(z) \neq 0$ in the interval, it follows that the term in braces must be zero. Thus,

$$r'(z) = b(z) + [a(z) + d(z)] r(z) + c(z) r^2(z), \quad 0 \leqslant z < z_1. \qquad \text{(V.2.4a)}$$

Comparing (V.2.1c) and (V.2.2) at $z = 0$, we find

$$r(0) = 0. \qquad\qquad \text{(V.2.4b)}$$

Accordingly, at $z = x_1$, the first zero of $u(z)$ to the right of $z = 0$, we have, from (V.2.1d) and (V.2.2),

$$r(x_1) = 0. \qquad\qquad \text{(V.2.5)}$$

Hence, in order to find the characteristic lengths we need only find the points where $r(z) = 0$. Suppose $v(z_1) = 0$. Since $u(z_1) = 0$ would violate the uniqueness hypothesis, it must be that $r(z)$ becomes infinite at z_1. To avoid obvious difficulties, we choose z_1', $0 < z_1' < z_1$ and observe that $r(z_1')$ exists. Next we introduce $s(z)$ by

$$v(z) = s(z)\, u(z). \qquad\qquad (V.2.6)$$

It is clear that $s(z)$ exists at z_1' and $s(z_1') = 1/r(z_1')$. Moreover, $s(z)$ is well-behaved at z_1 and, in fact, $s(z_1) = 0$. Indeed, by Sturm's theorem $s(z)$ exists until $z = x_1$. Finally, substitution of (V.2.6) into (V.2.1b), followed by use of (V.2.1a) and (V.2.6), gives

$$-s'(z) = c(z) + [a(z) + d(z)]\, s(z) + b(z)\, s^2(z), \qquad z_1' \leqslant z < x_1. \qquad (V.2.7)$$

As noted, the intial condition for $s(z)$ is

$$s(z_1') = 1/r(z_1'). \qquad\qquad (V.2.8)$$

To avoid the singularity of $s(z)$ at x_1, we choose x_1', $z_1 < x_1' < x_1$ and again define $r(z)$ as in (V.2.2). Equation (V.2.4a) is satisfied for $x_1' < z < z_2$, where z_2 is the next zero of $v(z)$. The process now continues in an obvious fashion. Figure 5-3 illustrates the process on a typical case. The solid line depicts the actual integration, and the dashed lines indicate the solutions of the $r(z)$ and $s(z)$ functions over the full interval.

It is apparent that exactly the same device furnishes the characteristic lengths for the system consisting of (V.2.1a, b) and

$$u(0) = 0, \qquad\qquad (V.2.9a)$$

$$v(x) = 0. \qquad\qquad (V.2.9b)$$

Indeed, the values $z_1, z_2, \ldots$ are precisely the characteristic lengths for this problem.

A convenient method for defining the points where the procedure switches from the r equation to the s equation and vice versa is to switch whenever $|r|$ or $|s|$ approaches unity. Thus, both r and s satisfy

$$-1 \leqslant r, \qquad s \leqslant 1,$$

on their respective domains of definition.

We may summarize our results in the following theorem.

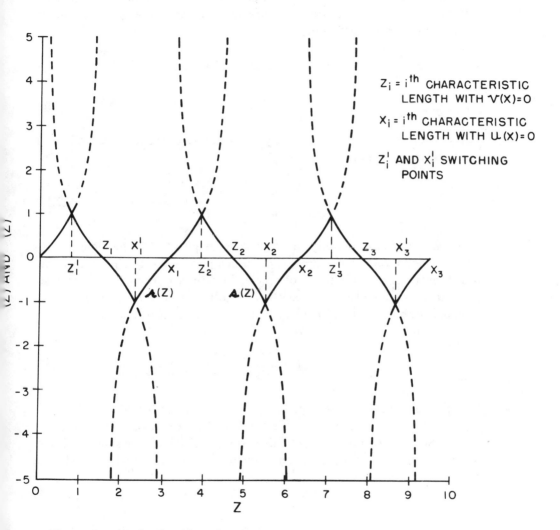

Figure 5-3.   Behavior of Typical *r* and *s* Functions

Theorem 1: If the coefficients a, b, c, and d are real and continuous on $0 \leqslant z \leqslant x \leqslant X$ so that the solutions of (V.2.1a, b) are oscillatory, then the location of the points or interval lengths x_i, where (V.2.1a, b) subject to (V.2.1c, d) has a nontrivial solution, are given by the zeros of the function $r(z)$. The location of the interval lengths z_i, where (V.2.1a, b) subject to (V.2.9a, b), has a nontrivial solution are given by the zeros of the function $s(z)$.

Corollary 1: Except at the points x_i and z_i, the functions $r(z)$ and $s(z)$ are related by

$$s(z) = 1/r(z). \tag{V.2.10}$$

The relationship (V.2.10) is actually a special case of the more general linear fractional transformation

$$s(z) = \frac{\alpha r(z) + \beta}{\gamma r(z) + \delta}, \qquad \Delta = \alpha\delta - \beta\gamma \neq 0, \tag{V.2.11}$$

where α, β, γ, and δ are functions of z. (See Chapter I.) The relation (V.2.10) corresponds to $\alpha=0$, $\beta=1$, $\gamma=1$, $\delta=0$, which gives $\Delta = -1$. Linear fractional transformations have a number of very interesting properties. From our viewpoint the most important is that the inverse transformation is also a linear fractional transformation. That is, the inverse of the transformation in (V.2.11) may be written as

$$r(z) = \frac{\hat{\alpha}s(z) + \hat{\beta}}{\hat{\gamma}s(z) + \hat{\delta}}. \tag{V.2.12}$$

It then follows that if $r(z)$ satisfies a Riccati equation, $s(z)$ does also.

We may use this fact to reduce the amount of work necessary in our computations. Suppose we are interested only in the first characteristic length (this was the case in the two examples given in section 1) for the boundary conditions

$$u(0) = 0, \quad v(x) = 0. \tag{V.2.13}$$

According to the above analysis, we want to find the first value of x so that $r(x) = +\infty$ or, equivalently, where $s(x) = 1/r(x) = 0$. This requires integrating the r equation up to some point z', say where $r(z') = 1$, and then switching to the s equation and continuing to integrate until $s(x) = 0$. In order to avoid switching equations during the course of integration, we may pick $\alpha=1$, $\beta=1$, $\gamma=1$, $\delta=+1$. Observe that, since $\Delta=2$, we have defined a nonsingular transformation. Now consider

$$s(z) = \frac{r(z) - 1}{r(z) + 1}. \tag{V.2.14}$$

Observe that at $z=0$, we have $s(0) = -1$; at $z=x$, we have $s(x) = +1$. Since $r(z)$ is a positive function on $0 \leqslant z \leqslant x$, it follows that $s(z)$ has no singularities on that interval. Thus we have reduced our problem to finding a value of x, where $s(x) = +1$ and $s(z)$ satisfies a Riccati equation with the initial condition $s(0) = -1$.

3 An Analytical Example

In order to illustrate the ideas of the previous section, let us consider a simple example which can be worked to completion in closed form. Consider

$$\frac{d^2 y}{dx^2} = - y(z), \tag{V.3.1a}$$

subject to

$$y(0) = 0, \tag{V.3.1b}$$

$$y(x) = 0. \tag{V.3.1c}$$

We wish to find the values of x so that (V.3.1) has a nontrivial solution. We first write (V.3.1a) as a system of two first-order equations by letting $u(z) = = y(z)$ and $v(z) = y'(z)$. Then (V.3.1a, b, c) become

$$\frac{du}{dz} = v(z), \tag{V.3.2a}$$

$$-\frac{dv}{dz} = u(z), \tag{V.3.2b}$$

subject to

$$u(0) = 0, \tag{V.3.2c}$$

$$u(x) = 0. \tag{V.3.2d}$$

The solution of this problem is easily obtained and is

$$u(z) = A \sin z, \quad A \text{ arbitrary}, \tag{V.3.3a}$$

$$v(z) = A \cos z, \tag{V.3.3b}$$

provided $\{x \in x_n : x_n = n\pi,\ n = 1, 2, \ldots\}$.

The differential equation satisfied by $r(z)$ becomes

$$r'(z) = 1 + r^2(z), \qquad r(0) = 0, \tag{V.3.4}$$

the solution of which is

$$r(z) = \tan z. \tag{V.3.5}$$

The zeros of the tangent function are given by the multiples of π, which agrees with the characteristic lengths given above. Recall that, except at a countable number of exceptional points, the functions $r(z)$ and $s(z)$ are related by

$$s(z) = 1/r(z) = \cotan z. \tag{V.3.6}$$

Hence, the singularities of the $r(z)$ function are the zeros of the $s(z)$ function. These zeros correspond to the characteristic lengths associated with (V.3.2a, b), subject to the boundary conditions

$$u(0) = 0, \tag{V.3.7a}$$

$$v(x) = 0. \tag{V.3.7b}$$

4   Modification of the Boundary Conditions

We now consider the calculation of the characteristic lengths for problems with general boundary conditions. Consider the system

$$\frac{du}{dz} = a(z)\, u(z) + b(z)\, v(z), \tag{V.4.1a}$$

$$-\frac{dv}{dz} = c(z)\, u(z) + d(z)\, v(z), \qquad 0 \leqslant z \leqslant x, \tag{V.4.1b}$$

subject to

$$\alpha_1 u(0) + \beta_1 v(0) + \gamma_1 u(x) + \delta_1 v(x) = 0, \tag{V.4.1c}$$

$$\alpha_2 u(0) + \beta_2 v(0) + \gamma_2 u(x) + \delta_2 v(x) = 0, \tag{V.4.1d}$$

where the conditions on the functions a, b, c, and d are as in section 3 and the constants α_i, β_i, γ_i, and δ_i ($i = 1, 2$) are such that there are a countable number of of x values such that (V.4.1) has a nontrivial solution. We have chosen the most general set of boundary conditions for a linear homogeneous problem. Notice that our previous conditions, (V.2.1c, d), (V.2.9a, b), and the important case of the periodic boundary conditons

$$u(0) = u(x), \tag{V.4.2a}$$

$$v(0) = v(x), \tag{V.4.2b}$$

are just special cases of (V.4.1c, d).

Since $u(0)$ may be nonzero and, in fact, unknown, we must modify our Riccati transformation of (V.2.2). The new transformation, which we shall refer to as the *generalized Riccati transformation*, is given by

$$u(z) = r_1(z) v(z) + r_2(z) u(0). \tag{V.4.3}$$

In addition, we introduce the *recovery transformation*

$$v(0) = q_1(z) v(z) + q_2(z) u(0). \tag{V.4.4}$$

These transformations can be physically motivated, but they simply represent a superposition of various contributions of u at z and v at $z=0$, respectively. In addition, these transformations can be easily modified to handle nonhomogeneous problems. Indeed, the method of invariant imbedding gives a complete theory of both nonhomogeneous and homogeneous problems.

We can derive the differential equations for r_1, r_2, q_1 and q_2 in a manner similar to that for the r function in section 2. Specifically, if we differentiate in (V.4.3) with respect to z and use (V.4.1a, b), followed by the use of (V.4.3) again, we obtain

$$\{r_1'(z) - b(z) - [a(z) + d(z)] r_1(z) - c(z) r_1^2(z)\} v(z)$$
$$+ \{r_2'(z) - a(z) r_2(z) - c(z) r_1(z) r_2(z)\} u(0) = 0 \tag{V.4.5}$$

We now argue that (V.4.5) must hold for all z, $0 \leqslant z \leqslant x$, and all $u(0)$. Thus, each term in braces in (V.4.5) must be zero. That is, the functions r_1 and r_2 satisfy as follows:

$$r_1'(z) = b(z) + [a(z) + d(z)] r_1(z) + c(z) r_1^2(z), \tag{V.4.6a}$$

$$r_2'(z) = [a(z) + c(z) r_1(z)] r_2(z). \tag{V.4.6b}$$

As for the initial conditions, a suitable set is obtained by evaluating (V.4.3) at $z=0$ and equating coefficients. Whence,

$$r_1(0) = 0, \qquad\qquad\qquad\qquad \text{(V.4.6c)}$$

$$r_2(0) = 1. \qquad\qquad\qquad\qquad \text{(V.4.6d)}$$

The differential equations for q_1 and q_2 are derived in a similar fashion and are

$$q_1'(z) = [d(z) + c(z) r_1(z)] q_1(z), \qquad\qquad \text{(V.4.7a)}$$

$$q_2'(z) = c(z) q_1(z) r_2(z), \qquad\qquad\qquad \text{(V.4.7b)}$$

with the initial conditions, obtained by evaluating (V.4.4) at $z=0$ and comparing coefficients,

$$q_1(0) = 1, \qquad\qquad\qquad\qquad \text{(V.4.7c)}$$

$$q_2(0) = 0. \qquad\qquad\qquad\qquad \text{(V.4.7d)}$$

Now evaluate (V.4.3) and (V.4.4) at $z=x$ and consider (V.4.1c, d), (V.4.3) and (V.4.4) as a system of four equations and four unknowns, which we write in matrix form as

$$
\begin{bmatrix}
\alpha_1 & \beta_1 & \gamma_1 & \delta_1 \\
\alpha_2 & \beta_2 & \gamma_2 & \delta_2 \\
r_2(x) & 0 & -1 & r_1(x) \\
q_2(x) & -1 & 0 & q_1(x)
\end{bmatrix}
\begin{bmatrix}
u(0) \\
v(0) \\
u(x) \\
v(x)
\end{bmatrix}
=
\begin{bmatrix}
0 \\
0 \\
0 \\
0
\end{bmatrix}. \qquad \text{(V.4.8)}
$$

A necessary and sufficient condition for the system in (V.4.8) to have a nontrivial solution is that the determinant of the matrix of coefficients be zero. Hence, for the determination of the characteristic lengths of (V.4.1a, b) subject to (V.4.1c, d), we need only determine the points, x, where the determinant of the matrix of coefficients in (V.4.8) is zero.

Since the r_1 function may become infinite during the course of integration, we must devise a method of integrating through the singularity. We do this by introducing the inverse Riccati transformation at $z=z'$,

$$v(z) = s_1(z) u(z) + s_2(z) v(z'), \qquad\qquad \text{(V.4.9)}$$

and the inverse recovery transformation,

$$u(z') = t_1(z) u(z) + t_2(z) v(z'). \qquad\qquad \text{(V.4.10)}$$

The differential equations satisfied by the s and t functions are

$$- s_1'(z) = c(z) + [a(z) + d(z)] s_1(z) + b(z) s_1^2(z), \qquad \text{(V.4.11a)}$$

$$- s_2'(z) = [d(z) + b(z) s_1(z)] s_2(z), \qquad \text{(V.4.11b)}$$

$$- t_1'(z) = [a(z) + b(z) s_1(z)] t_1(z), \qquad \text{(V.4.11c)}$$

$$- t_2'(z) = b(z) t_1(z) s_2(z), \qquad \text{(V.4.11d)}$$

with the initial conditions at $z=z'$

$$s_1(z') = 0, \qquad \text{(V.4.11e)}$$

$$s_2(z') = 1, \qquad \text{(V.4.11f)}$$

$$t_1(z') = 1, \qquad \text{(V.4.11g)}$$

$$t_2(z') = 0. \qquad \text{(V.4.11h)}$$

In order to use the transformations defined by (V.4.9) and (V.4.10), we must find the values of $u(z')$ and $v(z')$ in terms of $u(0)$ and $v(0)$. We may do this by evaluating (V.4.3) and (V.4.4) at $z=z'$ and solving for $u(z')$ and $v(z')$. We find

$$v(z') = \frac{1}{q_1(z')} v(0) - \frac{q_2(z')}{q_1(z')} u(0), \qquad \text{(V.4.12a)}$$

$$u(z') = \frac{r_1(z')}{q_1(z')} v(0) + \left[r_2(z') - \frac{r_1(z') q_2(z')}{q_1(z')} \right] u(0), \qquad \text{(V.4.12b)}$$

where the r and q functions are known from the integration of (V.4.6) and (V.4.7). (Notice that the function $q_1(z)$ satisfies a homogeneous differential equation (V.4.7a) subject to a nonzero initial condition (V.4.7c); hence, the function $q_1(z)$ is never zero and division by $q_1(z)$ in (V.4.12) is valid.) Substitution of these values into (V.4.9) and (V.4.10) yields the new transformations

$$v(z) = s_1(z) u(z) + s_2(z) [D_1 v(0) + C_1 u(0)], \qquad \text{(V.4.13a)}$$

$$[A_1 - C_1 t_2(z)] u(0) = t_1(z) u(z) + [D_1 t_2(z) - B_1] v(0), \qquad \text{(V.4.13b)}$$

where

$$D_1 = \frac{1}{q_1(z')}, \quad C_1 = -\frac{q_2(z')}{q_1(z')},$$

$$B_1 = \frac{r_1(z')}{q_1(z')}, \qquad A_1 = r_2(z') - \frac{r_1(z') q_2(z')}{q_1(z')}.$$

We may, of course, obtain a criterion for the determination of the characteristic lengths with these transformations as we did with (V.4.3) and (V.4.4). Evaluating (V.4.13) at $z = x$ and considering (V.4.13a, b) and (V.4.1.c, d) as four equations and four unknowns, we have

$$\begin{bmatrix} \alpha_1 & \beta_1 & \gamma_1 & \delta_1 \\ \alpha_2 & \beta_2 & \gamma_2 & \delta_2 \\ C_1 s_2(x) & D_1 s_2(x) & s_1(x) & -1 \\ C_1 t_2(x) - A_1 & D_1 t_2(x) - B_1 & t_1(x) & 0 \end{bmatrix} \begin{bmatrix} u(0) \\ v(0) \\ u(x) \\ v(x) \end{bmatrix} = \begin{bmatrix} 0 \\ 0 \\ 0 \\ 0 \end{bmatrix}.$$

$$\text{(V.4.14)}$$

Again we have that the characteristic lengths are the values of x so that the determinant of the matrix of coefficients in (V.4.14) is zero.

Theorem 2: If the coefficients a, b, c, and d are real and are such that the solutions of (V.4.1a, b) subject to the boundary conditions (V.4.1c, d) have a countable number of points, x_i, where the solutions are nonzero, these points correspond to the points where the determinant of the matrix of coefficients of (V.4.8) or (V.4.14) is zero.

If we wish to calculate additional characteristic lengths, we must continue to switch back and forth between the standard transformations (V.4.3, V.4.4) and the inverse transformations (V.4.9, V.4.10) each time one is about to become infinite. However, the transformations must be modified slightly. The new transformations are

$$u(z) = r_1(z) v(z) + r_2(z) u(z_i), \qquad \text{(V.4.15a)}$$

$$v(z_i) = q_1(z) v(z) + q_2(z) u(z_i), \quad i = 0, 2, 4, \ldots, \qquad \text{(V.4.15b)}$$

and

$$v(z) = s_1(z) u(z) + s_2(z) v(z_i), \qquad \text{(V.4.16a)}$$

$$u(z_i) = t_1(z) u(z) + t_2(z) v(z_i), \quad i = 1, 3, 5 \ldots. \qquad \text{(V.4.16b)}$$

where the z_i are the switching points and $z_0 = 0$. In each case we must relate the $u(z_i)$ and $v(z_i)$ to $u(0)$ and $v(0)$. Fortunately, there are some simple recursion formulas for these relations. These are simply a consequence of the linearity of the equations.

We write

$$u(z_i) = A_i u(0) + B_i v(0), \qquad\qquad \text{(V.4.17a)}$$

$$v(z_i) = C_i u(0) + D_i v(0), \qquad\qquad \text{(V.4.17b)}$$

where

$$A_i = a_i A_{i-1} + b_i C_{i-1},$$

$$B_i = a_i B_{i-1} + b_i D_{i-1},$$

$$C_i = c_i A_{i-1} + d_i C_{i-1},$$

$$D_i = c_i B_{i-1} + d_i D_{i-1},$$

with

$$A_0 = 1, \qquad B_0 = 0, \qquad C_0 = 0, \qquad D_0 = 1,$$

for i odd,

$$a_i = r_2(z_i) - r_1(z_i) \frac{q_2(z_i)}{q_1(z_i)},$$

$$b_i = \frac{r_1(z_i)}{q_1(z_i)},$$

$$c_i = \frac{q_2(z_i)}{q_1(z_i)},$$

$$d_i = \frac{1}{q_1(z_i)},$$

and, for i even,

$$a_i = \frac{1}{t_1(z_i)},$$

$$b_i = -\frac{t_2(z_i)}{t_1(z_i)},$$

$$c_i = \frac{s_1(z_i)}{t_1(z_i)},$$

$$d_i = s_2(z_i) - s_1(z_i) \frac{t_2(z_i)}{t_1(z_i)}.$$

These recursion relations are easily derived by successive substitution of

(V.4.17) into (V.4.15, V.4.16). In terms of these new transformations, the determinants (V.4.8) and (V.4.14) become

$$
\begin{bmatrix}
\alpha_1 & \beta_1 & \gamma_1 & \delta_1 \\
\alpha_2 & \beta_2 & \gamma_2 & \delta_2 \\
r_2(x) A_i & r_2(x) B_i & -1 & r_1(x) \\
q_2(x) A_i - C_i & q_2(x) B_i - D_i & 0 & q_1(x)
\end{bmatrix}
\begin{bmatrix}
u(0) \\
v(0) \\
u(x) \\
v(x)
\end{bmatrix}
=
\begin{bmatrix}
0 \\
0 \\
0 \\
0
\end{bmatrix}
$$

(V.4.18)

and

$$
\begin{bmatrix}
\alpha_1 & \beta_1 & \gamma_1 & \delta_1 \\
\alpha_2 & \beta_2 & \gamma_2 & \delta_2 \\
s_2(x) C_i & s_2(x) D_i & s_1(x) & -1 \\
t_2(x) C_i - A_i & t_2(x) D_i - B_i & t_1(x) & 0
\end{bmatrix}
\begin{bmatrix}
u(0) \\
v(0) \\
u(x) \\
v(x)
\end{bmatrix}
=
\begin{bmatrix}
0 \\
0 \\
0 \\
0
\end{bmatrix}.
$$

(V.4.19)

5 Further Analytical Examples

Now let us see how some of our earlier results and equations with periodic boundary conditions are just special cases of the general development given in the previous section.

Let A denote the matrix of coefficients in (V.4.8). For the boundary conditions (V.2.1c, d) the matrix A becomes

$$
A =
\begin{bmatrix}
1 & 0 & 0 & 0 \\
0 & 0 & 1 & 0 \\
r_2(x) & 0 & -1 & r_1(x) \\
q_2(x) & -1 & 0 & q_1(x)
\end{bmatrix}.
$$

Using elementary theory of determinants, the determinant of A can be easily shown to be

$$
|A| = -
\begin{vmatrix}
1 & 0 \\
-1 & r_1(x)
\end{vmatrix}
= -r_1(x).
$$

Hence, as in section 3, we wish to find the values of x so that $r_1(z) = 0$. Notice that, in this case, the only function which contributes to the calculation of the determinant is $r_1(z)$. Hence, the functions $r_2(z)$, $q_1(z)$, and $q_2(z)$ need not even be computed, thus justifying the reduced version of the transformations as given by (V.2.2) and (V.2.6).

The second set of boundary conditions to be considered is given by

$$\alpha_1 u(0) + \beta_1 v(0) = 0, \tag{V.5.1a}$$

$$\gamma_2 u(x) + \delta_2 v(x) = 0. \tag{V.5.1b}$$

We could, of course, use the development of the previous section to study (V.4.1) subject to (V.5.1a, b). However, since the boundary conditions are separated, the generalized transformations (V.4.3) and (V.4.4) are not necessary. That is, by substituting

$$u(z) = r(z) v(z) \tag{V.5.2}$$

into (V.5.1a), we obtain

$$\{\alpha_1 r(0) + \beta_1\} v(0) = 0. \tag{V.5.3}$$

If (V.5.1a) is not simply equivalent to the condition $v(0)=0$, we have

$$r(0) = -\beta_1/\alpha_1. \tag{V.5.4a}$$

Similarly, at $z=x$,

$$r(x) = -\delta_2/\gamma_2. \tag{V.5.4b}$$

The conditions on the s function may, of course, be written as

$$s(0) = -\alpha_1/\beta_1, \tag{V.5.5a}$$

$$s(x) = -\gamma_2/\delta_2. \tag{V.5.5b}$$

(Exceptional cases in which one or more of the parameters α_1, β_1, γ_2, and δ_2 vanish are easily handled.)

The same standard r and s substitutions obviously still yield (V.2.4a) and (V.2.7) when applied to (V.4.1a, b). Thus, the problem defined by (V.4.1a, b) and (V.5.1a, b) may be integrated by starting with (V.2.4a), subject to (V.5.4a) [or (V.2.7), subject to (V.5.4a)] and integrating until condition (V.5.4b) or (V.5.5b) occurs. The scheme of switching from r to s formulation—and vice versa—whenever one of the two functions is about to become singular must still be employed.

Finally, we may note that any nonsingular linear transformation

$$u(z) = A\tilde{u}(z) + B\tilde{v}(z), \tag{V.5.6a}$$

$$v(z) = C\tilde{u}(z) + D\tilde{v}(z), \qquad\qquad (\text{V.5.6b})$$

converts the problem (V.4.1a, b), (V.5.1a, b) into one of the same form. For numerical purposes, the new problem in $\tilde{u}$, $\tilde{v}$ functions may be of a simpler nature—for example, it may have less complicated coefficients or boundary conditions—than the original. For this reason, transformations such as (V.5.6) may be computationally useful.

The eigenvalue or eigenlength problem having periodic boundary conditions is one of the most interesting and difficult to handle with classical techniques. The primary reason for this is that each eigenvalue may be a twofold eigenvalue; that is, there may be two linearly independent eigenfunctions associated with each eigenvalue. For a second-order problem this can happen only with periodic boundary conditions.

Consider the equation

$$y'' = -\lambda y, \quad \lambda \text{ fixed}, \qquad\qquad (\text{V.5.7a})$$

subject to the periodic boundary conditions

$$y(0) = y(x), \qquad\qquad (\text{V.5.7b})$$

$$y'(0) = y'(x). \qquad\qquad (\text{V.5.7c})$$

In order to treat this problem by the method of invariant imbedding, we first convert (V.5.7a) into two first-order equations by letting $u(z)=y(z)$ and $v(z)=y'(z)$. Then u and v satisfy the differential equations

$$u'(z) = v(z), \qquad\qquad (\text{V.5.8a})$$

$$-v'(z) = \lambda u(z), \qquad\qquad (\text{V.5.8b})$$

subject to

$$u(0) = u(x), \qquad\qquad (\text{V.5.8c})$$

$$v(0) = v(x). \qquad\qquad (\text{V.5.8d})$$

The differential equations satisfied by the r and q functions are

$$r_1'(z) = 1 + \lambda r_1^2(z), \quad r_1(0) = 0, \qquad\qquad (\text{V.5.9a})$$

$$r_2'(z) = \lambda r_1(z) r_2(z), \quad r_2(0) = 1, \qquad\qquad (\text{V.5.9b})$$

$$q_1'(z) = \lambda r_1(z) q_1(z), \quad q_1(0) = 1, \tag{V.5.9c}$$

$$q_2'(z) = \lambda q_1(z) r_2(z), \quad q_2(0) = 0. \tag{V.5.9d}$$

Again, we can easily solve these equations in closed form as follows:

$$r_1(z) = \frac{1}{k} \tan kz, \tag{V.5.10a}$$

$$r_2(z) = \sec kz, \tag{V.5.10b}$$

$$q_1(z) = \sec kz, \tag{V.5.10c}$$

$$q_2(z) = k \tan kz, \tag{V.5.10d}$$

where $k = \sqrt{\lambda}$.

The matrix A is given by

$$A = \begin{bmatrix} 1 & 0 & -1 & 0 \\ 0 & 1 & 0 & -1 \\ r_2(x) & 0 & -1 & r_1(x) \\ q_2(x) & -1 & 0 & q_1(x) \end{bmatrix}.$$

We want to find the value of x so that $|A| = 0$. Upon expanding the determinant, we find that we want the value of x so that

$$1 - r_2(x) - q_1(x) + q_1(x) r_2(x) - r_1(x) q_2(x) = 0. \tag{V.5.11}$$

Substitution of (V.5.10) into (V.5.11) yields

$$0 = 1 - \cos kx = 2 \sin^2 \frac{k}{2} x, \tag{V.5.12}$$

a well-known result. Also, from (V.5.12) we see that there is a double root, thus each eigenvalue is a double eigenvalue.

6   Eigenfunctions

In the previous sections we were primarily interested in calculating the interval lengths so that systems of the form (V.4.1) have a nontrivial solution. However, many times we also need the nontrivial solution itself. We cannot, of course, determine this solution uniquely, since the eigenfunctions of any second-order system form a one-parameter family of solutions.

Recall that the generalized Riccati transformation and the recovery transformation are given by

$$u(z) = r_1(z) v(z) + r_2(z) u(0),$$ (V.6.1a)

$$v(0) = q_1(z) v(z) + q_2(z) u(0).$$ (V.6.1b)

Solving for $u(z)$ and $v(z)$, we obtain

$$u(z) = \left\{ \frac{r_2(z) q_1(z) - r_1(z) q_2(z)}{q_1(z)} \right\} u(0) + \frac{r_1(z)}{q_1(z)} v(0)$$ (V.6.2a)

$$v(z) = -\frac{q_2(z)}{q_1(z)} u(0) + \frac{1}{q_1(z)} v(0).$$ (V.6.2b)

From (V.6.2) we see that the solutions of the invariant imbedding equations, when combined properly, actually produce the general solution of (V.4.1). In In order to see that the coefficients of $u(0)$ and $v(0)$ in (V.6.2) do indeed generate independent solutions of (V.4.1), we define

$$w_1(z) = \frac{r_2(z) q_1(z) - r_1(z) q_2(z)}{q_1(z)},$$ (V.6.3a)

$$w_2(z) = \frac{r_1(z)}{q_1(z)},$$ (V.6.3b)

$$w_3(z) = \frac{q_2(z)}{q_1(z)},$$ (V.6.3c)

$$w_4(z) = \frac{1}{q_1(z)}.$$ (V.6.3d)

If we differentiate in the above equations and use (V.4.6) and (V.4.7), we find that the w functions satisfy the following initial-value problems

$$w_1'(z) = a(z) w_1(z) + b(z) w_3(z),$$ (V.6.4a)

$$-w_3'(z) = c(z) w_1(z) + d(z) w_3(z),$$ (V.6.4b)

$$w_1(0) = 1,$$ (V.6.4c)

$$w_3(0) = 0,$$ (V.6.4d)

and

$$w_2' (z) = a (z) w_2 (z) + b (z) w_4 (z), \qquad\qquad \text{(V.6.5a)}$$

$$- w_4' (z) = c (z) w_2 (z) + d (z) w_4 (z), \qquad\qquad \text{(V.6.5b)}$$

$$w_2 (0) = 0, \qquad\qquad \text{(V.6.5c)}$$

$$w_4 (0) = 1. \qquad\qquad \text{(V.6.5d)}$$

Thus the w functions generate two linearly independent solutions of (V.4.1). We may now write (V.6.2) in the form

$$\begin{pmatrix} u (z) \\ v (z) \end{pmatrix} = \begin{pmatrix} w_1 (z) & w_2 (z) \\ w_3 (z) & w_4 (z) \end{pmatrix} \begin{pmatrix} u (0) \\ v (0) \end{pmatrix} ; \qquad\qquad \text{(V.6.6)}$$

where the matrix

$$\Phi (z) = \begin{pmatrix} w_1 (z) & w_2 (z) \\ w_3 (z) & w_4 (z) \end{pmatrix}$$

is a fundamental matrix for the system (V.4.1.).

Hence, the eigenfunctions of the system (V.4.1) may be generated from the invariant imbedding equations by using (V.6.2).

Let us return to our example of section 3. Consider

$$\frac{du}{dz} = v (z), \qquad\qquad \text{(V.6.7a)}$$

$$- \frac{dv}{dz} = u (z), \qquad\qquad \text{(V.6.7b)}$$

subject to

$$u (0) = 0, \qquad\qquad \text{(V.6.7c)}$$

$$u (x) = 0. \qquad\qquad \text{(V.6.7d)}$$

If we are interested in the eigenfunctions for this problem, we must use (V.6.1) in the form

$$u (z) = r_1 (z) v (z), \qquad\qquad \text{(V.6.8a)}$$

$$v (0) = q_1 (z) v (z), \qquad\qquad \text{(V.6.8b)}$$

which yields

$$u (z) = \frac{r_1 (z)}{q_1 (z)} v (0), \qquad\qquad \text{(V.6.9a)}$$

$$v(z) = \frac{1}{q_1(z)} v(0).$$
(V.6.9b)

The differential equations satisfied by $r_1(z)$ and $q_1(z)$ are

$$r_1'(z) = 1 + r_1^2(z), \qquad r_1(0) = 0,$$
(V.6.10)

$$q_1'(z) = r_1(z) q_1(z), \qquad q_1(0) = 1.$$
(V.6.11)

The solutions, which are easily obtained, are

$$r_1(z) = \tan z,$$
(V.6.12)

$$q_1(z) = \sec z.$$
(V.6.13)

Substitution of (V.6.12) and (V.6.13) into (V.6.9) yields

$$u(z) = v(0) \sin z$$
(V.6.14a)

$$v(z) = v(0) \cos z,$$
(V.6.14b)

which agree with (V.3.3a, b). We must, of course, continue to use our inverse transformations, since $r_1(z)$ becomes infinite at $z = \pi/2$.

For simplicity let $z' = \pi/4$. Then (V.6.14) evaluated at $z = z'$ become

$$u(\pi/4) = \sqrt{2}/2 \, v(0),$$
(V.6.15a)

$$v(\pi/4) = \sqrt{2}/2 \, v(0).$$
(V.6.15b)

Since $v(z') \neq 0$, we must use the inverse transformation in the form

$$u(z) = s_1(z) u(z) + s_2(z) v(z'),$$
(V.6.16a)

$$u(z') = t_1(z) u(z) + t_2(z) v(z'), \qquad z' \leqslant z \leqslant z'' < x.$$
(V.6.16b)

The differential equations satisfied by $s_1(z)$, $s_2(z)$, $t_1(z)$ and $t_2(z)$ are

$$-s_1'(z) = 1 + s_1^2(z), \qquad s_1(z') = 0,$$
(V.6.17a)

$$-s_2'(z) = s_1(z) s_2(z), \qquad s_2(z') = 1,$$
(V.6.17b)

$$-t_1'(z) = s_1(z) t_1(z), \qquad t_1(z') = 1,$$
(V.6.17c)

$$-t_2'(z) = t_1(z) s_2(z), \qquad t_2(z') = 0.$$
(V.6.17d)

The solutions, which again are easily obtained, are

$$s_1(z) = \operatorname{ctn}(z + \pi/4), \tag{V.6.18a}$$

$$s_2(z) = \operatorname{csc}(z + \pi/4), \tag{V.6.18b}$$

$$t_1(z) = \operatorname{csc}(z + \pi/4), \tag{V.6.18c}$$

$$t_2(z) = \operatorname{ctn}(z + \pi/4). \tag{V.6.18d}$$

From (V.6.16) we have

$$u(z) = \frac{1}{t_1(z)}\{u(z') - t_2(z)\,v(z')\}, \tag{V.6.19a}$$

$$v(z) = \frac{s_1(z)}{t_1(z)}\{u(z') - t_2(z)\,v(z') + s_2(z)\,v(z')\}. \tag{V.6.19b}$$

Substituting (V.6.15) and (V.6.18) into (V.6.19), we find that

$$u(z) = v(0)\sin z, \tag{V.6.20a}$$

$$v(z) = v(0)\cos z, \quad z' \leqslant z \leqslant z'' < x, \tag{V.6.20b}$$

which agrees with (V.6.14) to give $u(z)$ and $v(z)$ over the interval $0 \leqslant z \leqslant z''$.

In order to extend the interval to $0 \leqslant z \leqslant x$, we must use the Riccati and recovery transformations in the form

$$u(z) = r_1(z)\,v(z) + r_2(z)\,u(z''), \tag{V.6.21a}$$

$$v(z'') = q_1(z)\,v(z) + q_2(z)\,u(z''). \tag{V.6.21b}$$

This form is necessary, since $u(z'') \neq 0$. Proceeding as before, we find

$$u(z) = v(0)\sin z, \tag{V.6.22a}$$

$$v(z) = v(0)\cos z, \quad z'' \leqslant z \leqslant x < z^2. \tag{V.6.22b}$$

Hence, we have calculated the eigenfunction over the interval $0 \leqslant z < z_2$. The interval can be easily extended by repeating the above process.

7   Results for Sturm-Liouville Systems

We now turn our attention to the calculation of eigenvalues and character-istic lengths for Sturm-Liouville problems of the form

$$\frac{d}{dz}\left[K(z, \lambda)\frac{du}{dz}\right] + G(z, \lambda)\, u(z) = 0, \qquad (V.7.1a)$$

subject to the boundary conditions

$$\alpha(\lambda)\, u(0) + \beta(\lambda)\, u'(0) = 0, \qquad (V.7.1b)$$

$$\gamma(\lambda)\, u(x) + \delta(\lambda)\, u'(x) = 0, \qquad (V.7.1c)$$

where K and G are continuous functions of z on $0 \leqslant z \leqslant x$ and K is positive and continuously differentiable. In addition, K and G are continuous and mono-tonic decreasing functions of λ for $\Lambda_1 \leqslant \lambda \leqslant \Lambda_2$. It is known that (V.7.1), for fixed x, has a discrete set of eigenvalues.

It follows from theorem 4 of Chapter I that the location of the characteristic lengths of (V.7.1) is a continuous monotonic decreasing function of λ. Hence, it has unique continuous inverse; that is, the eigenvalues of λ are a continuous function of x. In this sense, we state that the eigenvalue problem and eigenlength problem are equivalent. Thus, λ is an eigenvalue of (V.7.1) if x is a characteristic length.

Let

$$\frac{du}{dz} = \frac{1}{K(z, \lambda)}\, v(z) \qquad (V.7.2a)$$

so that

$$-\frac{dv}{dz} = G(z, \lambda)\, u(z) \qquad (V.7.2b)$$

This is a special case of (V.2.1a–d), and the corresponding r and s equations are

$$\frac{dr}{dz} = \frac{1}{K(z, \lambda)} + G(z, \lambda)\, r^2(z). \qquad (V.7.3)$$

$$-\frac{ds}{dz} = G(z, \lambda) + \frac{1}{K(z, \lambda)}\, s^2(z). \qquad (V.7.4)$$

The present formulation then yields

$$r(z) = \frac{u(z)}{v(z)} = \frac{u(z)}{K(z, \lambda) u'(z)}, \tag{V.7.5}$$

and

$$s(z) = \frac{K(z, \lambda) u'(z)}{u(z)}. \tag{V.7.6}$$

The boundary conditions are equivalent to

$$r(0) = -\frac{\beta(\lambda)}{K(0, \lambda) \alpha(\lambda)} = \frac{u(0)}{K(0, \lambda) u'(0)}, \tag{V.7.7}$$

$$r(x) = -\frac{\delta(\lambda)}{K(x, \lambda) \gamma(\lambda)}. \tag{V.7.8}$$

There are two principal ways of using our methods. Some problems, as discussed in section 1, are concerned with the characteristic lengths. More typically, one is given an interval $[0, b]$ and asked to find a λ so that a non-trivial solution which satisfies (V.7.1a–c) with $x=b$ exists. This amounts to solving the nonlinear equation in λ,

$$r(b) = -\frac{\delta(\lambda)}{K(b, \lambda) \gamma(\lambda)}. \tag{V.7.9}$$

The procedure, for computing eigenvalues, is to integrate (V.7.3), subject to the initial condition (V.7.7), out to $z-b$. If (V.7.9) is not satisfied, the value of λ is adjusted and the above process is repeated until a λ is found which satisfies (V.7.9). As discussed in section 4, an r function will normally become infinite in the cause of integration of (V.7.3). This difficulty is avoided by switching to the reciprocal s and using (V.7.4). (Indeed, this technique can be used to scale the numerical computations to whatever range is desired.)

One motivation for the use of this approach is evident in (V.7.5), (V.7.7), and (V.7.8). If one is interested only in eigenvalues, the values of u and u' are not required. It is only the ratio represented by (V.7.5) which is significant, and we would expect that computing it directly would yield greater accuracy than computing u and u' separately and then forming the ratio. Also, numerical integration routines are designed to adapt their step sizes in order to maintain accuracy. A direct approach requires a step size compatible with the poorest behavior of u and u', whereas this approach puts the step adjustment for accuracy exactly where it ought to be.

8 Singular Coefficients

In all of the previous sections we assumed that the coefficients were all continuous functions on the interval under question. This was done primarily to facilitate the presentation. However, there are many interesting and practical problems where one or more of the coefficients may be singular at one or, perhaps, both of the boundary points. We shall consider only regular singular points.

The solution of the differential equation in the neighborhood of a singular point frequently becomes large in magnitude, experiences rapid changes in magnitude, or exhibits other types of peculiar behavior. Thus, it is necessary to study the behavior at these points very carefully. Our analysis will proceed by studying the asymptotic behavior of the solutions near a singular point.

We shall consider equations both in their normal form and their selfadjoint form. The normal form will be given by

$$y''(z) + f(z)\, y'(z) + g(z)\, y(z) = 0, \qquad (V.8.1)$$

which we take to have a regular singular point at $z=0$, where

$$f(z) = \alpha/z + M(z), \qquad (V.8.2)$$

and

$$g(z) = N(z)/z^2, \qquad (V.8.3)$$

with $M(z)$ and $N(z)$ holomorphic and $N(z) = z^\gamma P(z)$ with $\gamma \geqslant 0$ and $P(0) \neq 0$.

The r equation of section 1 is easily seen to be

$$r'(z) = 1 + f(z)\, r(z) + g(z)\, r^2(z), \qquad (V.8.4)$$

or

$$r'(z) = 1 + \left[\frac{\alpha}{z} + M(z)\right] r(z) + z^{\gamma-2} P(z)\, r^2(z). \qquad (V.8.5)$$

In order for (V.8.5) to be numerically attractive, $r'(z)$ must be finite at $z=0$.

This requires the solutions of (V.8.1) be of the form $y = z^\beta Y(z)$, where $Y(0) \neq 0$ and $Y(z)$ is holomorphic. We shall consider several cases.

Case I. $[\beta > 0$, normal form, r equation]. For this case $y' = z^{\beta-1}T(z)$, where $T(z) = \beta Y(z) + z Y'(z)$ is holomorphic and $T(0) \neq 0$. Recall that, in this case, $r = y/y'$. Thus

$$r = z^\beta Y(z)/z^{\beta-1}T(z) = zY(z)/T(z). \qquad (V.8.6)$$

Substitution of (V.8.6) into (V.8.5) yields

$$r'(z) = 1 + [\alpha + zM(z)]\frac{Y(z)}{T(z)} + z^\gamma P(z)\frac{Y^2(z)}{T^2(z)}. \qquad (V.8.7)$$

Hence, r' remains bounded near $z=0$ for all values of α. Notice that this analysis is independent of γ, the order of the zero of $N(z)$ at $z=0$.

Case II. $[\beta = 0$, normal form, r equation]. For this case the second term of (V.8.4) is finite only when $\alpha = 0$ and the third term requires $\gamma \geqslant 2$.

Now consider the s transformation. The equation satisfied by the s function is

$$-s'(z) = z^{\gamma-2}P(z) + \left[\frac{\alpha}{z} + M(z)\right]s(z) + s^2(z). \qquad (V.8.8)$$

Case III. $[\beta > 0$, normal form, s equation]. Here, $s = y'/y = T(z)/zY(z)$. Hence, (V.8.8) becomes

$$-s'(z) = z^{\gamma-2}P(z) + \left[\frac{\alpha}{z} + M(z)\right]\frac{T(z)}{zY(z)} + \frac{T^2(z)}{z^2Y^2(z)}. \qquad (V.8.9)$$

For this case s' is unbounded at $z=0$.

Case IV. $[\beta > 0$, normal form, s equation]. If $y'(0) = 0$, then s' remains finite at $z=0$ for all values of α, provided that $\gamma \geqslant 2$.

We may put (V.8.1) into self-adjoint form by multiplying by $\exp \int^z f(t)\, dt = z^\alpha Q(z)$, where $Q(z) = \exp \int^z M(t)\, dt$. We obtain

$$[z^\alpha Q(z)\, y'(z)]' + z^{\alpha-2}Q(z)\, N(z)\, y(z) = 0,$$

with $Q(z) > 0$ for $z \geqslant 0$ and $N(z) = z^\gamma P(z)$, where $P(0) \neq 0$ and $\gamma \geqslant 0$. If we proceed as in section 7, we let

$$\frac{du}{dz} = \frac{1}{z^{\alpha}Q(z)} \, v(z), \tag{V.8.10a}$$

$$-\frac{dv}{dz} = z^{\alpha+\gamma-2}Q(z) \, P(z) \, u(z), \tag{V.8.10b}$$

where $u(z) = y(z)$ and $v(z) = z^{\alpha}Q(z) \, y'(z)$. We now set

$$r(z) = \frac{u(z)}{v(z)} = \frac{y(z)}{z^{\alpha}Q(z) \, y'(z)}, \tag{V.8.11}$$

and

$$s(z) = \frac{v(z)}{u(z)} = \frac{z^{\alpha}Q(z) \, y'(z)}{y(z)}. \tag{V.8.12}$$

Then $r(z)$ and $s(z)$ satisfy the following differential equations:

$$r' = \frac{1}{z^{\alpha}Q(z)} + z^{\alpha+\gamma-2}Q(z) \, P(z) \, r^2(z), \tag{V.8.13}$$

and

$$-s' = z^{\alpha+\gamma-2}Q(z) \, P(z) + \frac{1}{z^{\alpha}Q(z)} \, s^2(z). \tag{V.8.14}$$

Let us first consider the r equation.

Case V. $[\beta > 0$, self-adjoint form, r equation$]$. Again, we have $y = z^{\beta}Y(z)$ and $y' = z^{\beta-1}T(z)$. Using (V.8.11), we may rewrite (V.8.13) as

$$r' = \frac{1}{z^{\alpha}Q(z)} + z^{-\alpha+\gamma} \frac{P(z) \, Y^2(z)}{Q(z) \, T^2(z)}. \tag{V.8.15}$$

For r' to remain bounded at $z = 0$, we must have $\alpha \leqslant 0$ with $\gamma \geqslant 0$.

Case VI. $[\beta = 0$, self-adjoint form, r equation$]$. We have

$$r' = \frac{1}{z^{\alpha}Q(z)} + z^{-\alpha+\gamma-2} \frac{P(z) \, Y^2(z)}{Q(z) \, T^2(z)}. \tag{V.8.16}$$

which is bounded for $\alpha \leqslant 0$ and $\gamma \geqslant 2 + \alpha$.

Let us now consider the s transformation as given by (V.8.12).

Case VII. $[\beta>0$, self-adjoint form, s equation]. Using $y=z^\beta Y(z)$ and $y'=z^{\beta-1}T(z)$, we may rewrite (V.8.14) as

$$- s' = z^{\alpha+\gamma-2}P(z)\,Q(z) + z^{\alpha-2}Q(z)\,\frac{T^2(z)}{Y^2(z)}, \qquad \text{(V.8.17)}$$

which is finite for all $\alpha \geqslant 2$ and $\gamma \geqslant 0$. If $y'(0)=0$, then s' is finite for $\alpha \geqslant 0$ and $\gamma \geqslant 2-\alpha$.

Case VIII. $[\beta>0$, self-adjoint form, s equation]. In this case, (V.8.14) becomes

$$- s' = z^{\alpha+\gamma-2}P(z)\,Q(z) + z^{\alpha-2}Q(z)\,\frac{T^2(z)}{Y^2(z)}, \qquad \text{(V.8.18)}$$

which is finite for $\alpha \geqslant 0$ and $\gamma \geqslant 2-\alpha$.
We summarize our results in the following:

Theorem 3: If there is a solution of the form $y(z)=z^\beta Y(z)$, $Y(0)\neq0$, $\beta\geqslant0$ of the normal equation

$$y''(z) + f(z)\,y'(z) + g(z)\,y(z) = 0, \qquad \text{(V.8.19)}$$

where $f(z)=\alpha/z+M(z)$ and $g(z)=N(z)/z^2$ with $M(z)$ and $N(z)$ holomorphic and $N(z)=z^\gamma P(z)$, $P(0)\neq0$, then (a) r', as given by (V.8.5), is finite in a neighborhood of $z=0$, provided that one of the following conditions holds:

    (i) $\beta>0$, $\gamma\geqslant0$ and any finite value of α,

    (ii) $\beta=0$, $\gamma\geqslant2$ and $\alpha=0$.

(b) The term s', as given by (V.8.8), is finite in a neighborhood of $z=0$, provided that the following condition holds:

    (i) $\beta=0$, $y'(0)=0$, $\gamma\geqslant2$ and any finite value of α.

Theorem 4: If there is a solution of the form $y=z^\beta Y(z)$, $Y(0)\neq0$, $\beta\geqslant0$ of the equation

$$[z^\alpha Q(z)\,y'(z)]' + z^{\alpha-2}N(z)\,Q(z)\,y(z) = 0, \qquad \text{(V.8.20)}$$

where $Q(z)>0$ and $N(z)=z^\gamma P(z)$, $P(0)\neq 0$, then (a) r', as given by (V.8.13), is finite in a neighborhood of $z=0$, provided that one of the following conditions holds:

$$\text{(i)} \quad \beta>0, \quad \alpha\leqslant 0 \quad \text{and} \quad \gamma\geqslant 0$$

$$\text{(ii)} \quad \beta=0, \quad \alpha\leqslant 0 \quad \text{and} \quad \gamma\geqslant 2+\alpha.$$

(b) The term s', as given by (V.8.14), is finite in a neighborhood at $z=0$, provided that one of the following conditions holds:

$$\text{(i)} \quad \beta>0, \quad \alpha\geqslant 0$$

$$\text{(ii)} \quad \beta>0, \quad y'(0)=0, \quad \alpha\geqslant 0, \quad \gamma\geqslant 2-\alpha$$

$$\text{(iii)} \quad \beta=0, \quad \alpha\geqslant 0, \quad \gamma\geqslant 2-\alpha$$

$$\text{(iv)} \quad \beta=0, \quad y'(0)=0, \quad \alpha\leqslant 0 \quad \text{and} \quad \gamma\geqslant 2-\alpha.$$

9   Singular Intervals

The second type of singularity which we shall discuss occurs when the interval is the semi-infinite interval, $0\leqslant z<+\infty$, or the infinite interval, $-\infty<z<+\infty$. Both types occur frequently in quantum mechanical investigations and can be difficult to handle.

We shall consider equations of the form

$$y''(z)+[\lambda-q(z)]\,y(z)=0, \qquad (V.9.1)$$

where $q(z)$ is continuous on $(-\infty,\infty)$ and $q(z)\rightarrow+\infty$ as $x\rightarrow+\infty$ and as $x\rightarrow-\infty$. This form is chosen because it occurs quite frequently in practical problems. Physicists refer to it as the "bound case." As always, we assume that there exists a countable number of eigenvalues of the equation. This means that each eigenfunction has a finite number of zeros and approaches zero as z approaches infinity.

For the semi-infinite case, the procedure is straightforward. The invariant imbedding equations are all initial-valued and, hence, can be integrated from $z=0$ in the direction of increasing z. As the invariant imbedding equations are passing through each zero of the eigenfunctions, the switching process must be performed as in the case of a finite interval. Recall that in our process of finding eigenvalues we must pick a value of λ and continue adjusting it until the boundary condition at infinity is satisfied. Since computers perform only finite arithmetic, we will never be able to satisfy this requirement exactly. However, since the solutions of (V.9.1) for any set of initial conditions become unbounded as z approaches infinity except for λ exactly an eigenvalue, both the r equation and the s become unbounded as well. This gives a very sensitive bounding criterion in determining the eigenvalue.

A few numerical results for both the semi-infinite and finite interval cases are presented in section 10.

10 · Connections with Other Methods

A number of researchers have used phase functions to compute eigenvalues of Sturm-Liouville problems such as (V.7.1a, b). They use the Prüfer transformation

$$u(z) = \zeta(z) \sin \theta(z) \tag{V.10.1a}$$

$$K(z, \lambda) u'(z) = \zeta(z) \cos \theta(z). \tag{V.10.1b}$$

It is easily verified that the phase function, $\theta(z)$, satisfies

$$\theta'(z) = \frac{1}{K(z, \lambda)} \cos^2 \theta(z) + G(z, \lambda) \sin^2 \theta(z). \tag{V.10.2}$$

Since

$$\tan \theta(z) = \frac{u(z)}{K(z, \lambda) u'(z)}, \tag{V.10.3}$$

the boundary conditions may be expressed as

$$\theta(0) = \zeta \tag{V.10.4a}$$

$$\theta(x) = \eta + n\pi, \qquad n = 0, 1, 2, ...), \tag{V.10.4b}$$

where

$$\tan \zeta = - \frac{\beta_1(\lambda)}{K(0, \lambda) \alpha_1(\lambda)}, \tag{V.10.5a}$$

$$\tan \eta = - \frac{\delta_2(\lambda)}{K(x, \lambda) \gamma_2(\lambda)}. \tag{V.10.5b}$$

Comparing (V.10.3) and (V.7.5), we see that

$$r(z) = \tan \theta(z) \tag{V.10.6}$$

and, similarly,

$$s(z) = \cot \theta(z). \tag{V.10.7}$$

The invariant imbedding or Riccati approach is obviously closely related to the phase function approach for a single second order equation. It is not clear how one would generalize the phase approach to a pair or more of first order equations; however, as we shall see in Chapter VIII, this can be done in a natural way for the Riccati approach. The r and s equations are cheaper to evaluate because it is not necessary to evaluate sines and cosines. The only relative disadvantage is that one must occasionally switch between equations. This is minor because of the extremely simple relationship and, moreover, the benefits of scaling accrue. Both processes involve finding an x where an appropriate boundary condition is satisfied. This amounts to finding where a function assumes a given value. Both are satisfied at the same x. How well x is determined depends on the slope of the functions at x. Now,

$$r'(z) = [\sec^2 \theta(z)] \, \theta'(z). \tag{V.10.8}$$

Thus, at x,

$$|r'(z)| = [\sec^2 \eta] \, |\theta'(x)| \geqslant |\theta'(x)|. \tag{V.10.9}$$

Evidently the x at which the boundary condition is satisfied is always determined for the $r - s$ procedure at least as well as for the phase method; the larger $\sec^2 \eta$, the greater the advantage. Figure 1a of Bailey's paper shows that the $\theta(z)$ may have flat spots where the characteristic lengths are poorly determined, and the observation just made becomes important.

11   Numerical Examples

We shall consider several examples to illustrate the power of the method. It should be emphasized that a single program of only moderate sophistication was used in all of these applications. All computations were performed on a CDC-3600 computer, using a standard fourth order Runge-Kutta integration scheme. We have not tried to obtain high accuracy, since our goal was to demonstrate feasibility.

Example 1

As we have remarked, the method is particularly powerful when we are primarily interested in characteristic lengths. Such a problem, as we have seen, arises in the study of particle transport in a simple one-dimensional rod. When each end of the rod adjoins a reflecting material, the governing equation is

$$y'' + \lambda \sigma^2 y = 0, \qquad \text{(V.11.1a)}$$

with the boundary conditions

$$\beta_1 y'(0) - \sigma y(0) = 0, \qquad \text{(V.11.1b)}$$

$$y'(x) - \sigma \beta_2 y(x) = 0. \qquad \text{(V.11.1c)}$$

Here, β_1 and β_2 are the so-called albedo numbers for the reflecting materials, and σ is the constant macroscopic cross section of the rod. Hence, x is the critical length of the rod when $\lambda = 1$ is an eigenvalue of (V.11.1) associated with a nonnegative function $y(x)$. Further details of the physical model can be found in section 1.

This problem can be resolved analytically as a check on our method. For $\beta_1 = 0.2$, $\beta_2 = 0.2$, and $\sigma = 10.0$, we found relative errors of approximately 6.7×10^{-4} when a step size of 10^{-2} was used, and errors of 2.3×10^{-4} when a step size of 5×10^{-3} was used. These errors are consistent with the accuracy of the Runge-Kutta integration of the differential equation.

Example 2

One of the simplest eigenvalue problems is

$$y'' + \lambda y = 0, \qquad \text{(V.11.2a)}$$

$$y(-1) = 0, \qquad \text{(V.11.2b)}$$

$$y(x) = 0. \qquad \text{(V.11.2c)}$$

A great many numerical methods resolve this problem satisfactorily. We obtained accuracy to five significant figures in the calculation of the characteristic lengths and characteristic functions. The characteristic lengths are displayed in Figure 5-4 to demonstrate how one may determine the eigenvalues for various x.

Example 3

The nonlinear eigenvalue problem

$$y'' + (\lambda + \lambda^2 z^2) y = 0, \qquad \text{(V.11.3a)}$$

$$y(-1) = 0, \qquad \text{(V.11.3b)}$$

$$y(x) = 0, \qquad \text{(V.11.3c)}$$

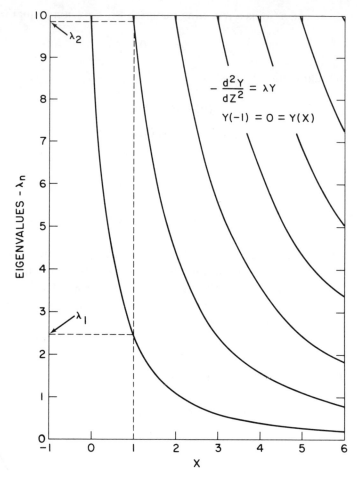

Figure 5-4.   Eigenvalue-Eigenlength Curves for $-y''=\lambda y$, $y(-1)=0=y(x)$

which may be considered as a perturbation of (V.11.2), is a different matter. Most of the methods for solving (V.11.3) require considerable modification to deal with this problem. Our scheme requires no change. The characteristic curves are displayed in Figure 5-5. The first three eigenvalues for $x=1$ were found to be

$$\lambda_1 = 1.9517$$

$$\lambda_2 = 4.2861$$

$$\lambda_3 = 7.5459.$$

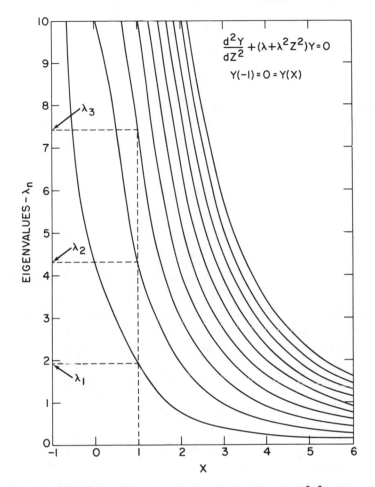

Figure 5-5.   Eigenvalue-Eigenlength Curves for $y'' + (\lambda + \lambda^2 z^2)y = 0$, $y(-1) = 0 = y(x)$

In discussing this problem, Collatz gives bounds for the first eigenvalue:

$$1.811 \leqslant \lambda_1 \leqslant 1.965.$$

Our computed value is obviously consistent with this bound.

Example 4

In section 5 we analytically analyzed the periodic problem

$$y'' = -\lambda y, \qquad\qquad\qquad \text{(V.11.4a)}$$

$$y(0) = y(x), \qquad\qquad\qquad \text{(V.11.4b)}$$

$$y'(0) = y'(x). \qquad\qquad\qquad \text{(V.11.4c)}$$

The numerical results for the calculation of the characteristic lengths and characteristic functions agreed with the analytical results to five significant figures.

Example 5

We shall now discuss some singular problems. The anharmonic oscillator problem is

$$y'' + (\lambda - z^2 - \varepsilon z^4)\, y = 0, \qquad\qquad \text{(V.11.5a)}$$

$$y(-\infty) = 0, \qquad\qquad\qquad \text{(V.11.5b)}$$

$$y(+\infty) = 0. \qquad\qquad\qquad \text{(V.11.5c)}$$

Bazley and Fox have calculated both upper and lower bounds for the first eigenvalues corresponding to even eigenfunctions for $\varepsilon = 0.0, 0.1, 0.2, ..., 1.0$. Here, we have considered only the most difficult case, $\varepsilon = 1.0$, using two methods.

The first procedure was to replace the boundary conditions by

$$y(-5) = 0 = y(+5). \qquad\qquad\qquad \text{(V.11.5d)}$$

We checked in a few cases to determine whether -5 was a fair representation for $-\infty$ by using, instead, $-6, -7, -8, -9, -10$. The computed eigenlengths agreed to the accuracy of the machine. The results, given in Table V.5-11, are consistent with the bounds of Bazley and Fox.

The second procedure was to use the known symmetry properties of the eigenfunctions to consider (V.11.5a) on $(0, \infty)$. In the even case, we may use

$$y(0) = 1, \quad y(+\infty) = 0, \qquad\qquad \text{(V.11.5e)}$$

and, in the odd case,

$$y(0) = 0, \quad y(+\infty) = 0. \qquad\qquad \text{(V.11.5f)}$$

Figure 5-6 shows the very interesting behavior of the r and s functions for the calculation of the eighth eigenvalue, an odd symmetry case. We have plotted the r and s functions for $\lambda = 34.6408$ and $\lambda = 34.6409$ versus the independent variable

z. For $z \leqslant 3.12$, the r functions are almost identical. Since $\lambda = 34.6408$ is less than the correct value of λ, the r function drops drastically to $-\infty$. Accordingly, since $\lambda = 34.6409$ is greater than the correct value of λ, the r function approaches zero

TABLE V.11-1.   *Eigenvalues for Anharmonic Oscillator* $\varepsilon = 1.0$

	Our Results Rounded to Five Figures	Lower Bounds (Bazley and Fox)	Upper Bounds (Bazley and Fox)
λ_1	1.3923	1.390301	1.393371
λ_2	4.6488		
λ_3	8.6551	8.330586	8.689663
λ_4	13.157		
λ_5	18.058	15.62953	21.74203
λ_6	23.297		
λ_7	28.835	21.0000	59.22833
λ_8	34.641		
λ_9	40.690	25.0000	167.6966
λ_{10}	46.950		

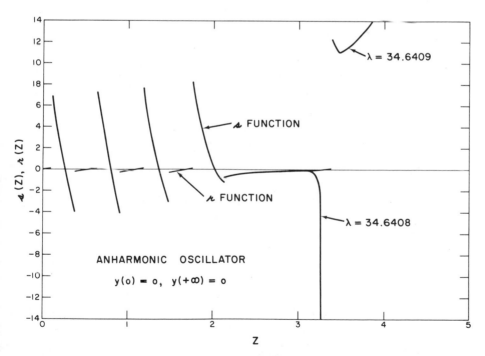

Figure 5-6.   Comparison of the r and s Functions of the Anharmonic Oscillator for Two Different Values of λ

and becomes slightly positive. We then switch to the s function and it goes to $+\infty$. Because of this sensitive behavior of the r and s functions near $+\infty$, we are able to compute the eigenvalues very accurately.

Example 6

The eigenvalue problem

$$\frac{d}{dz}\left\{ z \frac{dy}{dz} \right\} + 4\lambda z\left(1 - z^2\right) y\left(z\right) = 0, \tag{V.11.6a}$$

$$y\left(0\right) \quad \text{finite}, \tag{V.11.6b}$$

$$y'\left(1\right) = 0, \tag{V.11.6c}$$

arises in heat and mass transfer. We use it as an example because there is a mild singularity at $z=0$ and because there have been extensive computations which integrate (V.11.6a) directly. The physical origin of the problem, by virtue of symmetry, requires $y'(0)=0$, which is the condition we use in the form

$$s\left(0\right) = 0 \text{ since } y'\left(z\right) = \frac{s\left(z\right)}{z} \, y\left(z\right).$$

The use of our method is straightforward. We computed the first five eigenvalues and obtained agreement with those of Dranoff to a relative error of less than 3×10^{-5}.

Example 7

Latzko's equation

$$\frac{d}{dz}\left\{ \left(1 - z^7\right) \frac{dy\left(z\right)}{dz} \right\} + \lambda z^7 y\left(z\right) = 0, \tag{V.11.7a}$$

$$y\left(0\right) = 0, \tag{V.11.7b}$$

$$y\left(1\right) = \text{finite} \tag{V.11.7c}$$

has been studied by expansion in Legendre and Jacobi polynomials, as well as direct integration using Taylor's series. We discuss it for this reason and because of the strong singularity at $z=1$. From the series expansion at $z=1$, an appropriate boundary condition is found to be $y'(1)=0$. Table V.11-2 compares the straightforward use of our method with various other approaches for the first three eigenvalues.

TABLE V.11-2.   *Eigenvalues for Latzko's Equation*

	Our Method	Latzko	Fettis	Durfee
λ_1	8.728	8.71	8.72798	8.72747
λ_2	152.45	164.4	152.8	152.42
λ_3	435.2	1700.0	462.5	435.06

12   Exercises

1. Assume that $r(z)$ satisfies the Riccati equation

$$r'(z) = 1 + r^2(z),$$

with the initial condition $r(0)=0$. Let

$$s(z) = \frac{r(z) - 1}{r(z) + 1}.$$

Derive the differential equation satisfied by $s(z)$ and show that $s(z)$ is well defined on $0 \leqslant z \leqslant x$, where x is the first value of z so that $r(x) = +\infty$. Find x.

2. Derive the recursion relations given by (V.4.17).

3. Obtain the self-adjoint form of (V.8.1).

4. Discuss a technique for solving (V.9.1) on $(-\infty, +\infty)$.

5. Verify (V.10.2).

6. Verify that (V.5.10a–d) are solutions of (V.5.9a–d).

13   Bibliographical Discussion

Section 1

The best discussion of eigenvalue problems is found in

L. Collatz, *"Eigenwertaufgaben mit Technischen Anwendungen,"* Akademische Verlagsgesellschaft, Leipzig, 1963.

Section 2

The material of this section first appeared in

M. R. Scott, L. F. Shampine and G. M. Wing, "Invariant Imbedding and the Calculation of Eigenvalue for Sturm-Liouville Systems," *Computing* **4** (1969), 10–23.

Section 5

For a discussion of the conditions on K and G for (5.1) to have a discrete spectrum see

E. L. Ince, *Ordinary Differential Equations*, Dove Publ., Inc., New York, 1944.

Section 8

Here, we extend the development given in

D. O. Banks and G. J. Kurowski, "Computation of Eigenvalues of Singular Sturm-Liouville Systems," *Math. Comp.* **22** (1968), 304–310.

Section 10

The phase function approach has been the standard approach to eigenvalue problems for many years. There have been a number of excellent papers concerning the numerical aspects of the approach. See

P. B. Bailey, "Sturm-Liouville Eigenvalues via a Phase Function," *J. SIAM Appl. Math.* **14** (1966), 242–249.

D. O. Banks and G. J. Kurowski, "Computation of Eigenvalues of Singular Sturm-Liouville Systems," *Math. Comp.* **22** (1968), 304–310.

E. A. Coddington and N. Levinson, "Theory of Ordinary Differential Equations," *McGraw-Hill Book Co.*, New York, 1955.

M. Godart, "An Iterative Method for the Solution of Eigenvalue Problems," *Math. Comp.* **20** (1966), 399–406.

N. Wax, "On a Phase Method for Treating Sturm-Liouville Equations and Problems," *J. SIAM Appl. Math.* **9** (1961), 215–232.

Section 11

The numerical examples of this section are, for the most part, nontrivial problems which have occurred in physical applications. The intent is to make comparisons of the method with other techniques for problems with which the practicing engineer and mathematician may come in contact.

Example 1

R. E. Bellman and R. E. Kalaba, "Transport Theory and Invariant Imbedding," *Proc. Symposium in Appl. Math.* **11** (1959), 206–218.

Example 2

L. Collatz, *Eigenwertaufgaben mit Technischen Anwendungen*, Akademische Verlagsgesellschaft, Leipzig, 1963.

Example 3

L. Collatz, "Monotonicity and Related Methods in Non-Linear Differential Equations," a chapter in *Numerical Solutions of Non-Linear Differential Equations*, Donald Greenspan, ed., John Wiley and Sons, Inc., New York, 1966.

Example 4

L. Collatz, *Eigenwertaufgaben mit Technischen Anwendungen*, Akademische Verlagsgesellschaft, Leipzig, 1963.

Example 5

N. W. Bazley and D. W. Fox, "Lower Bounds for Eigenvalues of Schrödinger's Equations," *Phys. Review* **124** (1961), 483–492.

Example 6

J. S. Dranoff, "An Eigenvalue Problem Arising in Mass and Heat Transfer Studies," *Math. Comp.* **15** (1961), 403–409.

Example 7

W. H. Durfee, "Heat Flow in a Fluid with Eddying Flow," *J. Aero. Sci.* **23** (1956), 188–189.

H. E. Fettis, "On the Eigenvalues of Latzko's Differential Equation," *Z. Angew. Math. Mech.* **37** (1957), 398–399.

H. Latzko, "Warmübergang an einem Turbulenten Flüssigkeits- oder Gasstrom," *Z. Angew. Math. Mech.* **1** (1921), 268–290.

LINEAR BOUNDARY-VALUE PROBLEMS – INHOMOGENEOUS EQUATIONS REVISITED

1 Introduction

The purpose of this chapter is to summarize and correlate the results of Chapters IV and V in terms of the classical alternative theorem of section 5 of Chapter I. In section 2 we present this theorem and in section 3 we discuss the invariant imbedding formulation on intervals greater than the first characteristic length. Several numerical examples are presented in section 4.

2 Alternative Theorem

We consider the differential equations

$$u'(z, x) = a(z) u(z, x) + b(z) v(z, x) + e(z), \qquad \text{(VI.2.1a)}$$

$$- v'(z, x) = c(z) u(z, x) + d(z) v(z, x) + f(z), \qquad \text{(VI.2.1b)}$$

subject to the two-point boundary conditions

$$\alpha_1 u(0, x) + \beta_1 v(0, x) + \gamma_1 u(x, x) + \delta_1 v(x, x) = \eta_1, \qquad \text{(VI.2.1c)}$$

$$\alpha_2 u(0, x) + \beta_2 v(0, x) + \gamma_2 u(x, x) + \delta_2 v(x, x) = \eta_2, \qquad \text{(VI.2.1d)}$$

where $0 \leqslant z \leqslant x \leqslant X$. We assume that the coefficients a, b, c, and d are real-valued and continuous on $[0, X]$. The coefficients α_i, β_i, γ_i, and δ_i $(i = 1, 2)$ are assumed to be real-valued constants.

Recall that we may introduce the Riccati and recovery transformations by the relations

$$u(z, x) = r_1(z) v(z, x) + r_2(z) u(0, x) + r_3(z), \qquad \text{(VI.2.2a)}$$

$$v(0, x) = q_1(z) v(z, x) + q_2(z) u(0, x) + q_3(z). \qquad \text{(VI.2.2b)}$$

The r and q functions satisfy a system of weakly coupled first-order differential equations with initial conditions. (See Chapter 4, section 5). If we evaluate (VI.2.1c, d) and (VI.2.2a, b) at $z = x$ and view the resulting equations as a system of four equations for the four unknowns, $u(0, x)$, $v(0, x)$, $u(x, x)$, and $v(x, x)$, we have in matrix form

$$
\begin{bmatrix}
\alpha_1 & \beta_1 & \gamma_1 & \delta_1 \\
\alpha_2 & \beta_2 & \gamma_2 & \delta_2 \\
r_2(x) & 0 & -1 & r_1(x) \\
q_2(x) & -1 & 0 & q_1(x)
\end{bmatrix}
\begin{bmatrix}
u(0, x) \\
v(0, x) \\
u(x, x) \\
v(x, x)
\end{bmatrix}
=
\begin{bmatrix}
\eta_1 \\
\eta_2 \\
-r_3(x) \\
-q_3(x)
\end{bmatrix}.
\qquad \text{(VI.2.3)}
$$

From the theory of linear algebraic equations, we know that the foregoing system has a unique solution as long as the determinant of the matrix of coefficients does not vanish. Also, we know that if the foregoing system does have a zero determinant, then, in general, there is no solution to the inhomogeneous system (VI.2.3) and there are an infinite number of solutions to the homogeneous system.

If the boundary conditions are homogeneous, then $\eta_1 = \eta_2 = 0$. If the source terms $e(z)$ and $f(z)$ are identically zero, then $r_3(z)$ and $q_3(z)$ are identically zero (see exercise 1). In this case the system of equations (VI.2.3) is a homogeneous system, and the problem is to find the values of z so that the determinant of the matrix of coefficients is zero. These points yield the characteristic lengths so that the homogeneous problem associated with (VI.2.1) has nontrivial solutions. We then have the results summarized in the theorem.

Theorem 1: The inhomogeneous boundary-value problem (VI.2.1) has a unique solution for a given x if, and only if, the determinant of the matrix of coefficients in (VI.2.3) is nonzero.

Corollary 1: The points x where the determinant of the matrix of coefficients in (VI.2.3) is zero are the characteristic lengths for the homogeneous problem associated with (VI.2.1).

There are many research problems still open to the interested reader in the foregoing discussion. We have scratched only the surface.

3   Solutions on Intervals Greater than First Characteristic Length

One of the major complaints of the invariant imbedding method is: "The invariant imbedding equations frequently go to infinity before the desired interval length is attained." While it is possible for this to occur, it should be of no great concern. As indicated by the theorem of the previous section and in Chapter V, the points of infinity of the imbedding equations yield information concerning characteristic lengths of the associated homogeneous problems. The purpose of this section is to develop an algorithm for solving linear boundary-value problem on intervals greater than the first characteristic length.
   Consider

$$u'(z) = a(z) u(z) + b(z) v(z) + e(z), \qquad \text{(VI.3.1a)}$$

$$-v'(z) = c(z) u(z) + d(z) v(z) + f(z), \qquad \text{(VI.3.1b)}$$

$$\alpha_1 u(0) + \beta_1 v(0) + \gamma_1 u(x) + \delta_1 v(x) = \eta_1, \qquad \text{(VI.3.1c)}$$

$$\alpha_2 u(0) + \beta_2 v(0) + \gamma_2 u(x) + \delta_2 v(x) = \eta_2, \qquad \text{(VI.3.1d)}$$

where x is chosen greater than the first characteristic length for the homogeneous equation and the homogeneous boundary conditions

$$u(0) = 0, \quad v(x) = 0. \qquad \text{(VI.3.2)}$$

The process for avoiding the singularities of the imbedding equations for the above problem is similar to that for homogeneous problems as presented in section 4 of Chapter V.
   For the interval $0 \leqslant z \leqslant z_1 < x_1$, we use the Riccati and recovery transformations in the form

$$u(z) = r_1(z) v(z) + r_2(z) u(z_0) + r_3(z), \qquad \text{(VI.3.3a)}$$

$$v(z_0) = q_1(z) v(z) + q_2(z) u(z_0) + q_3(z), \qquad \text{(VI.3.3b)}$$

where $z_0 = 0$. In order to avoid the singularity at x_1, we switch to the inverse transformations at $z = z_1$

$$v(z) = s_1(z) u(z) + s_2(z) v(z_1) + s_3(z), \qquad \text{(VI.3.4a)}$$

$$u(z_1) = t_1(z) u(z) + t_2(z) v(z_1) + t_3(z). \qquad \text{(VI.3.4b)}$$

Since the boundary conditions of (VI.3.1) are specified only at $z=0$ and $z=x$, we must express $u(z_1)$ and $v(z_1)$ in terms of $u(z_0)$ and $v(z_0)$. This can be accomplished by evaluating (VI.3.3) at $z=z_1$, solving for $u(z_1)$ and $v(z_1)$ and then substituting the result into (VI.3.4). We obtain

$$u(z) = \frac{1}{t_1(z)} u(z_1) - \frac{t_2(z)}{t_1(z)} v(z_1) - \frac{t_3(z)}{t_1(z)} \tag{VI.3.5}$$

$$v(z) = s_1(z) u(z) + s_2(z) v(z_1) + s_3(z), \tag{VI.3.6}$$

where

$$v(z_1) = \frac{1}{q_1(z_1)} v(z_0) - \frac{q_2(z_1)}{q_1(z_1)} u(z_0) - \frac{q_3(z_1)}{q_1(z_1)},$$

$$u(z_1) = \frac{r_1(z_1)}{q_1(z_1)} v(z_0) + \left\{ r_2(z_1) - r_1(z_1) \frac{q_2(z_1)}{q_1(z_1)} \right\} u(z_0) +$$

$$+ \left\{ r_3(z_1) - r_1(z_1) \frac{q_3(z_1)}{q_1(z_1)} \right\}$$

If the interval length x contains several characteristic lengths, then the switching must be performed each time a characteristic length is approached. Fortunately, there are some recursion formulas for these relations, which are simply a consequence of the linearity of the equations.

We write the transformations as

$$u(z) = r_1(z) u(z) + r_2(z) u(z_i) + r_3(z), \tag{VI.3.7a}$$

$$v(z_i) = q_1(z) v(z) + q_2(z) u(z_i) + q_3(z), \quad i = 0, 2, 4, \dots, \tag{VI.3.7b}$$

and

$$v(z) = s_1(z) u(z) + s_2(z) v(z_i) + s_3(z), \tag{VI.3.8a}$$

$$u(z_i) = t_1(z) u(z) + t_2(z) v(z_i) + t_3(z), \quad i = 1, 3, 5, \dots, \tag{VI.3.8b}$$

where the z_i are the switching points and $z_0 = 0$. The recursion relations are

$$u(z_i) = A_i u(0) + B_i v(0) + E_i, \tag{VI.3.9a}$$

$$v(z_i) = C_i u(0) + D_i v(0) + F_i, \tag{VI.3.9b}$$

where

$$A_i = a_i A_{i-1} + b_i C_{i-1},$$

$$B_i = a_i B_{i-1} + b_i D_{i-1},$$

$$C_i = c_i A_{i-1} + d_i C_{i-1},$$

$$D_i = c_i B_{i-1} + d_i D_{i-1},$$

$$E_i = a_i E_{i-1} + b_i F_{i-1} + e_i,$$

$$F_i = c_i E_{i-1} + d_i F_{i-1} + f_i,$$

with $A_0 = 1$, $B_0 = 0$, $C_0 = 0$, $D_0 = 1$, $E_0 = 0$, $F_0 = 0$, and for i odd

$$a_i = r_2(z_i) - r_1(z_i) \frac{q_2(z_i)}{q_1(z_i)},$$

$$b_i = \frac{r_1(z_i)}{q_1(z_i)},$$

$$c_i = \frac{q_2(z_i)}{q_1(z_i)},$$

$$d_i = \frac{1}{q_1(z_i)},$$

$$e_i = r_3(z_i) - r_1(z_i) \frac{q_3(z_i)}{q_1(z_i)},$$

$$f_i = -\frac{q_3(z_i)}{q_1(z_i)},$$

and for i even,

$$a_i = \frac{1}{t_1(z_i)},$$

$$b_i = -\frac{t_2(z_i)}{t_1(z_i)},$$

$$c_i = \frac{s_1(z_i)}{t_1(z_i)},$$

$$d_i = s_2(z_i) - s_1(z_i) \frac{t_2(z_i)}{t_1(z_i)},$$

$$e_i = -\frac{t_3(z_i)}{t_1(z_i)},$$

$$f_i = s_3(z_i) - s_1(z_i)\frac{t_3(z_i)}{t_1(z_i)},$$

These recursion relations are easily derived by successive substitution of (VI.3.9) into (VI.3.7) and (VI.3.8).

4 Numerical Examples

We shall consider two examples where the interval length is longer than the first characteristic length. In all of the other examples of this book a standard fourth-order Runge-Kutta integration scheme with fixed stepsize has been employed. However, the examples considered in this section, especially the second, are best integrated with a code using a variable stepsize. The integration scheme used was a fourth-order Runge-Kutta procedure employing England's estimation of local truncation error to adjust the stepsize.

Example 1

Consider

$$y'' + \pi^2 y = 0, \tag{VI.4.1}$$

$$y(0) = 0, \quad y(x) = 1. \tag{VI.4.2}$$

This problem for the boundary conditions

$$y(0) = 0, \quad y'(x) = 0, \tag{VI.4.3}$$

has characteristic lengths, located at integer multiples of 1/2. The numerical experiment was performed with $x=5$ and the switching process was performed each time $|r_1(z)|$ and $|s_1(z)|$ exceeded 1. At least six significant figures were obtained in the solution on this interval.

Example 2

The second experiment was performed on the equation

$$y''(z) + (3\cot z + 2\tan z)\,y'(z) + 7y(z) = 0, \tag{VI.4.4}$$

$$y(30°) = 0, \quad y(60°) = 5. \tag{VI.4.5}$$

This problem is quite difficult to solve because the solution has a very sharp spike occurring approximately at $30.65°$. The magnitude of the solution at its peak is approximately 283. In this case the variable stepsize integration routine is a necessity because of the steep slope early in the solution and rather slowly varying slope after the peak is reached. The switch from the Riccati and recovery transformations to the inverse transformations occurred when $r_1(z) > 100$. The results are displayed in Table VI.4-1 and agreed to six significant figures with calculations using another technique.

TABLE VI.4-1.   *Invariant Imbedding Solution of*
$y'' + (3 \cot an z + 2 \tan z)y' + 0.7y = 0$,
$y(30°) = 0$, $y(60°) = 5$ *Using Single-Precision Arithmetic on EMR-6130*

z	u(z)	v(z)
30.00	0.	0.189644E+04
30.05	0.812229E+02	0.137942E+04
30.10	0.140250E+03	0.100132E+04
30.15	0.183041E+03	0.724720E+03
30.20	0.213954E+03	0.522290E+03
30.25	0.236171E+03	0.374093E+03
30.30	0.252022E+03	0.265566E+03
30.35	0.263210E+03	0.186067E+03
30.40	0.270982E+03	0.127818E+03
30.45	0.276250E+03	0.851322E+02
30.50	0.276894E+03	0.538489E+02
30.55	0.281774E+03	0.309231E+02
30.60	0.282878E+03	0.141254E+02
30.65	0.283261E+03	0.182294E+01
30.70	0.283116E+03	−0.718078E+01
30.75	0.282583E+03	−0.137628E+02
30.80	0.281769E+03	−0.185664E+02
30.85	0.280748E+03	−0.220633E+02
30.90	0.279578E+03	−0.245598E+02
30.95	0.278300E+03	−0.264303E+02
31.00	0.276944E+03	−0.277415E+02
31.50	0.262024E+03	−0.302237E+02
32.00	0.247189E+03	−0.290366E+02
36.00	0.151227E+03	−0.193422E+02
40.00	0.890707E+02	−0.121522E+02
44.00	0.509215E+02	−0.726841E+01
48.00	0.285301E+02	−0.417827E+01
52.00	0.158417E+02	−0.233204E+01
56.00	0.882555E+01	−0.127729E+01
60.00	0.500000E+01	−0.693964E+00

5 Exercises

1. Show that, if $e(z)$ and $f(z)$ are identically zero, $r_3(z)$ and $q_3(z)$ are identically zero.

2. Derive the recursion relations given in (VI.3.9).

3. Show that the mean of the coefficient of $y'(z)$ in (VI.4.4) is 5.24548729. The minimum and maximum values are, respectively, 4.89920 and 6.35085.

4. It is then reasonable to expect that the solution of

$$y'' + my' + 0.7y = 0,$$

$$y(30°) = 0, \quad y(60°) = 5,$$

where m is 5.24548729 is a good approximation of (VI.4.4, VI.4.5). Show this by solving the foregoing system and comparing with the results of Table VI.6-1.

5. Consider the initial-value problem

$$y'' + my' + 0.7y = 0,$$

$$y(30°) = 0, \quad y'(30°) = a.$$

Show that the solution of this problem is

$$y(\varphi) = Ae^{c_1\varphi} + Be^{c_2\varphi}$$

where

$$c_1 = -5.10845968$$

$$c_2 = -0.13702761$$

and

$$A = \frac{ae^{-c_1 30}}{c_1 - c_2}$$

$$B = \frac{-ae^{-c_2 30}}{c_1 - c_2}.$$

Since, for large φ, the solution varies only slightly with large changes in a, none of the classical initial-value techniques for solving the boundary-value problem (VI.4.4, VI.4.5) would be appropriate.

6   Bibliographical Discussion

Section 3

Other methods have been suggested for avoiding the singularities of the Riccati equations. See

J. L. Casti, R. E. Kalaba, and M. R. Scott, "A Proposal for the Calculation of Characteristic Functions for Certain Differential and Integral Operators Via Initial-Value Procedures," *J. Math. Anal. Appl.* **41** (1973), 1–13.

VII

LINEAR INITIAL-VALUE PROBLEMS

1 Introduction

In the previous chapters we were concerned with converting certain unstable boundary-value problems into stable initial-value problems. The natural question to ask is: If the invariant imbedding can transform certain unstable boundary-value problems into stable initial-value problems, can it also transform unstable initial-value problems into stable initial-value problems? The answer is in the affirmative with certain qualifications.

There are several different cases in which classical initial-value methods are directly applicable, undesirable, or must be slightly modified. The direct forward integration is stable when the growth of the wanted solution is the same as or greater than the growth of the dominant complementary solution. If it happens that both solutions of the homogeneous equation rise more rapidly than or decrease more slowly than the true solution, backward integration may be more appropriate, since both complementary solutions then decrease more rapidly or rise more slowly than the solution which is sought and, hence, usually cause no trouble. The case that usually causes the most trouble is where the solution growth is between the complementary solutions. Here, neither forward nor backward integration is stable, and very special techniques may be required.

Other excellent approaches for solving unstable initial-value problems which use boundary-value techniques have been developed. Although the problems of stability are, in many instances, eliminated in the boundary-value techniques, the questions of existence and uniqueness may be troublesome. Recall, for example, from Chapter I that the intial-value problem

$$y'' + \pi^2 y = 0, \tag{VII.1.1}$$

$$y(0) = 0, \tag{VII.1.2}$$

$$y'(0) = 1, \tag{VII.1.3}$$

has a unique solution over any finite interval, whereas the boundary-value problems defined by (VII.1.1), (VII.1.2) and

$$y(1) = 0 \qquad\qquad\qquad \text{(VII.1.4)}$$

or

$$y(1) = 1 \qquad\qquad\qquad \text{(VII.1.5)}$$

or

$$y'(1) = 1 \qquad\qquad\qquad \text{(VII.1.6)}$$

have, respectively, an infinite number of solutions, no solution, and a unique solution. Thus, it is clear that, when using boundary value techniques, care must be exercised.

In this chapter we shall show how the generalized Riccati transformation may be used to convert certain unstable initial-value problems into an equivalent initial-value problem which often is quite stable. In addition, we shall discuss the applicability of backward integration to certain types of unstable problems.

2   Formal Results

We consider two first-order linear differential equations of the form

$$u'(z) = a(z) u(z) + b(z) v(z) + e(z), \qquad \text{(VII.2.1)}$$

$$-v'(z) = c(z) u(z) + d(z) v(z) + f(z), \qquad \text{(VII.2.2)}$$

with the initial conditions

$$u(0) = \alpha, \quad v(0) = \beta, \qquad\qquad \text{(VII.2.3)}$$

where, for simplicity, α and β are real constants and the functions a, b, c, d, e, and f are real and continuous on $[0, \infty)$.

Since $u(0)$ and $v(0)$ are both specified, the Riccati and recovery transformations become

$$u(z) = r_1(z) v(z) + \alpha r_2(z) + r_3(z), \qquad \text{(VII.2.4)}$$

$$\beta = q_1(z) v(z) + \alpha q_2(z) + q_3(z). \qquad \text{(VII.2.5)}$$

The r and q functions are the same as those introduced in section 5 of Chapter IV. The differential equations, repeated here for completeness, satisfied by the r and q functions are

$$r_1'(z) = b(z) + [a(z) + d(z)] r_1(z) + c(z) r_1^2(z), \qquad \text{(VII.2.6a)}$$

$$r_2'(z) = [a(z) + c(z) r_1(z)] r_2(z), \qquad \text{(VII.2.6b)}$$

$$r_3'(z) = [a(z) + c(z) r_1(z)] r_3(z) + f(z) r_1(z) + e(z), \qquad \text{(VII.2.6c)}$$

$$q_1'(z) = [d(z) + c(z) r_1(z)] q_1(z), \qquad \text{(VII.2.6d)}$$

$$q_2'(z) = c(z) q_1(z) r_2(z), \qquad \text{(VII.2.6e)}$$

$$q_3'(z) = [c(z) r_3(z) + f(z)] q_1(z), \qquad \text{(VII.2.6f)}$$

with the initial conditions

$$\begin{aligned}
r_1(0) &= 0, & q_1(0) &= 1, \\
r_2(0) &= 1, & q_2(0) &= 0, \qquad \text{(VII.2.7)} \\
r_3(0) &= 0, & q_3(0) &= 0.
\end{aligned}$$

Thus, our algorithm involves solving (VII.2.6) subject to (VII.2.7) and then using (VII.2.4) and (VII.2.5) to obtain $u(z)$ and $v(z)$. At first sight it appears that we are going to a great deal of effort; however, as we shall see in the section on numerical examples, the initial-value problems defined by (VII.2.6) and (VII.2.7) are quite stable for many problems which are unstable as classical initial-value problems. In addition, we see that the solutions of (VII.2.6), (VII.2.7) are independent of α and β. Thus, once we have solved (VII.2.6), (VII.2.7), we can solve (VII.2.1) and (VII.2.2) for various values of α and β by using (VII.2.4) and (VII.2.5).

Depending on the form of the equations in (VII.2.1), (VII.2.2), we may wish to use the inverse transformations

$$v(z) = s_1(z) u(z) + \beta s_2(z) + s_3(z), \qquad \text{(VII.2.8)}$$

$$\alpha = t_1(z) u(z) + \beta t_2(z) + t_3(z). \qquad \text{(VII.2.9)}$$

The differential equations satisfied by the s and t functions are

$$-s_1'(z) = c(z) + [a(z) + d(z)] s_1(z) + b(z) s_1^2(z), \qquad \text{(VII.2.10a)}$$

$$-s_2'(z) = [d(z) + b(z) s_1(z)] s_2(z), \qquad \text{(VII.2.10b)}$$

$$- s_3'(z) = [d(z) + b(z) s_1(z)] s_3(z) + e(z) s_1(z) + f(z), \quad \text{(VII.2.10c)}$$

$$- t_1'(z) = [a(z) + b(z) s_1(z)] t_1(z), \quad \text{(VII.2.10d)}$$

$$- t_2'(z) = b(z) t_1(z) s_2(z), \quad \text{(VII.2.10e)}$$

$$- t_3'(z) = [b(z) s_3(z) + e(z)] t_1(z), \quad \text{(VII.2.10f)}$$

with the initial conditions

$$s_1(0) = 0, \quad s_2(0) = 1, \quad s_3(0) = 0, \quad \text{(VII.2.11a)}$$

$$t_1(0) = 1, \quad t_2(0) = 0, \quad t_3(0) = 0. \quad \text{(VII.2.11b)}$$

3   Numerical Examples

In order to illustrate the advantage of the invariant imbedding approach over the classical techniques, we shall consider two examples of unstable initial-value problems. The observations concerning the numerical solution of (VII.2.6), subject to (VII.2.7), and the computation of (VII.2.4) and (VII.2.5) given in section 14 of Chapter IV apply here as well. In particular, the use of the method of successive starts is advisable.

Example 1

Our first example is

$$y''(z) - (z^2 - 1) y(z) = 0, \quad \text{(VII.3.1)}$$

$$y(0) = 1, \quad y'(0) = 0. \quad \text{(VII.3.2)}$$

The general solution of (VII.3.1) is given by

$$y(z) = A e^{-z^2/2} + B e^{-z^2/2} \int_0^z e^{t^2} \, dt \quad \text{(VII.3.3)}$$

and, with the initial conditions (VII.3.2), the desired solution is

$$y(z) = e^{-z^2/2}, \quad \text{(VII.3.4)}$$

which is clearly the minimal solution. Thus, standard initial-value techniques will have difficulty in accurately approximating the solution over an interval of any appreciable length. Also, since $\exp(-z^2/2)$ is dominant with decreasing z, a backward integration scheme should be feasible.

In order to apply the invariant imbedding approach, we first convert (VII.3.1) into the two first-order equations

$$u'(z) = v(z), \tag{VII.3.5}$$

$$-v'(z) = -(z^2 - 1)\,u(z), \tag{VII.3.6}$$

$$u(0) = 1, \quad v(0) = 0, \tag{VII.3.7}$$

where $u(z) = y(z)$ and $v(z) = y'(z)$. There are an infinite number of ways to convert a second-order equation into a pair of first-order equations. It is possible that some ways may be better than others for numerical computation.

The form of the initial conditions in (VII.3.2) and the coefficients in (VII.3.5) and (VII.3.6) indicate that the relations (VII.2.17) and (VII.2.18) are most convenient and, since $e(z)$ and $f(z)$ are identically zero, the functions $s_3(z)$ and $t_3(z)$ are identically zero. In addition, since $\beta = v(0) = 0$, there is no need to solve for $s_2(z)$ and $t_2(z)$. Thus, our algorithm for this example becomes

$$-s_1'(z) = -(z^2 - 1) + s_1^2(z), \quad s_1(0) = 0, \tag{VII.3.8}$$

$$-t_1'(z) = s_1(z)\,t_1(z), \quad t_1(0) = 1, \tag{VII.3.9}$$

or

$$-s_1'(z) = -(z^2 - 1) + s_1^2(z), \quad s_1(0) = 0, \tag{VII.3.10}$$

$$u'(z) = s_1(z)\,u(z), \quad u(0) = 1. \tag{VII.3.11}$$

For the backward integration scheme, take

$$y(Z) = 1, \quad y'(Z) = 0, \quad Z \geqslant 10, \tag{VII.3.12}$$

and integrate backward from $z = Z$ to $z = 0.0$ and normalize all tabulated values by $y(0)$. Repeat for $Z' > Z$ and compare answers.

The original problem (VII.3.1, VII.3.2), the second of the invariant imbedding algorithms (VII.3.10, VII.3.11), and the backward integration scheme were solved on a CDC-6600 using a fourth-order Runge-Kutta integration scheme with $z = 0.005$. The results, displayed in Table VII.3-1, clearly demonstrate the advantage of the invariant imbedding and the backward integration schemes.

TABLE VII.3-1.   *Comparison of Invariant Imbedding, Forward and Backward Integration for*
$y'' - (z^2 - 1)y = 0, y(0) = 1, y'(0) = 0$

z	$y(z)$ Exact	Invariant Imbedding	Classical Initial Value	Backward Integration
0	1.0	1.0	1.0	1.0
1	6.065307$-$01	6.065307$-$01	6.065307$-$01	6.065306$-$01
2	1.353353$-$01	1.353353$-$01	1.353353$-$01	1.353353$-$01
3	1.110900$-$02	1.110900$-$02	1.110900$-$02	1.110900$-$02
4	3.354626$-$04	3.354626$-$04	3.354648$-$04	3.354627$-$04
5	3.726653$-$06	3.726653$-$06	3.879808$-$06	3.726653$-$06
6	1.522998$-$08	1.522998$-$08	3.103631$-$05	1.522998$-$08
7	2.289735$-$11	2.289735$-$11	1.761642$-$02	2.289736$-$11
8	1.266417$-$14	1.266417$-$14	2.780004$-$01	1.266417$-$14
9	2.576757$-$18	2.576758$-$18	1.212432$+$05	2.576759$-$18
10	1.928750$-$22	1.928752$-$22	1.456040$+$09	1.928756$-$22

Example 2

Our last example is

$$y''(z) - 11y'(z) - 12y(z) + 22e^z = 0, \qquad \text{(VII.3.13)}$$

$$y(0) = 1, \quad y'(0) = 1, \qquad \text{(VII.3.14)}$$

which we write in the form

$$u'(z) = v(z), \qquad \text{(VII.3.15)}$$

$$-v'(z) = -12u(z) - 11v(z) + 22e^z, \qquad \text{(VII.3.16)}$$

$$u(0) = 1, \quad v(0) = 1. \qquad \text{(VII.3.17)}$$

The general solution is given by

$$y(z) = Ae^{-z} + Be^{12z} + e^z, \qquad \text{(VII.3.18)}$$

and the initial conditions yield the solution

$$y(z) = e^z. \qquad \text{(VII.3.19)}$$

This problem is difficult to handle with any initial-value technique on the original equation, since the exact solution, $\exp(z)$, grows at a rate between the solutions of the homogeneous equation, $\exp(-z)$ and $\exp(12z)$. With forward integration, $\exp(12z)$ eventually takes over; whereas $\exp(-z)$ grows on backward integration and overtakes $\exp(z)$, which decreases on backward integration.

Our relations in this case will be (VII.2.4) and (VII.2.5), and this leads to the equations defined by (VII.2.6, VII.2.7). The results of soving (VII.3.13, VII.3.14) and (VII.2.6, VII.2.7) are displayed in Table VII.3-2. Although both techniques

TABLE VII.3-2.   *Comparison of Invariant Imbedding and Forward Integration for* $y'' - 11y' - 12y + 22e^z = 0,\ y(0) = 1,\ y'(0) = 1$

z	$y(z)$ Exact	Single Precision Invariant Imbedding	Double Precision Invariant Imbedding	Single Precision Classical Initial Value	Double Precision Classical Initial Value
0.0	1.0	1.0	1.0	1.0	1.0
0.2	1.221403	1.221403	1.221403	1.221403	1.221403
0.4	1.491825	1.491825	1.491825	1.491825	1.491825
0.6	1.822119	1.822119	1.822119	1.822119	1.822119
0.8	2.225541	2.225541	2.225541	2.225539	2.225539
1.0	2.718282	2.718282	2.718282	2.718258	2.718258
1.2	3.320117	3.320118	3.320117	3.319851	3.319851
1.4	4.055100	4.055211	4.055200	4.052266	4.052266
1.6	4.953031	4.953166	4.953035	4.920689	4.920691
1.8	6.049647	6.051146	6.049675	5.693119	5.693140
2.0	7.389056	7.407488	7.389360	3.458983	3.459217

are beginning to show deterioration, the invariant imbedding is considerably more stable. This particular problem would probably be best solved by using an implicit finite difference boundary technique.

4   Exercises

1. There are several ways to generate two linearly independent solutions of a second-order equation. Consider the equation

$$y'' = y. \qquad (*)$$

Show that u_1 and u_2 are linearly independent solutions of (*), where u_1 satisfies (*) and the initial conditions

$$u_1(0) = 1, \quad u_1'(0) = 0,$$

and u_2 satisfies (∗) and the initial conditions

$$u_2(0) = 0, \quad u_2'(0) = 1.$$

2. Show that v_1 and v_2 are linearly independent solutions of (∗), where v_1 satisfies (∗) and the initial conditions

$$v_1(0) = 1, \quad v_1'(0) = 1,$$

and v_2 satisfies (∗) and the initial conditions

$$v_2(0) = 1, \quad v_2'(0) = -1.$$

3. In using the principle of superposition to find the solution of

$$y'' = y,$$
$$y(0) = 0, \quad y(1) = e^{-1},$$

would it be best to use the functions u_1 and u_2 or the functions v_1 and v_2?

4. Formulate the backward integration scheme for inhomogeneous problems.

5. Formulate the method of successive starts for initial-value problems.

6. Formulate the method of Kagiwada and Kalaba for initial-value problems.

5 Bibliographical Discussion

Section 1

For excellent discussions of the application of boundary-value techniques to initial-value problems see

L. Fox and A. R. Mitchell, "Boundary-Value Techniques for the Numerical Solution of Initial-Value Problems in Ordinary Differential Equations," *Quart. Journ. Mech. and Appl. Math.* **10** (1957), 232–243.

D. Greenspan, "Approximate Solution of Initial-Value Problems for Ordinary Differential Equations by Boundary-Value Techniques," *J. Math. and Phys. Sci.* **1** (1967), 261–274.

R. A. Usmani, "Boundary-Value Techniques for the Numerical Solution of Certain Initial-Value Problems in Ordinary Differential Equations," *J. Assoc. Comp. Mach.* **13** (1966), 287–295.

Section 3

The only known application of imbedding techniques to initial-value problems appears in

M. R. Scott, "Numerical Solution of Unstable Initial-Value Problems by Invariant Imbedding", *The Computer Journal* **13** (1970), 397–399.

VIII
SYSTEMS OF EQUATIONS

1 Introduction

The invariant imbedding technique can be applied to n-th order systems. One particular benefit of the approach is its amenability to numerical calculations. With the introduction of large high-speed computers, the calculation of characteristic lengths, values, and functions for large systems of equations began to attract more attention. This has been particularly evident in the areas of structural mechanics, quantum mechanics, and nuclear physics. It soon became apparent that many of the classical techniques, such as the phase function approach and the implicit finite difference approach, lost some of their appeal because they were either difficult to generalize to systems or required large blocks of storage. The invariant imbedding approach is attractive on both counts. The generalization of the previous discussion to systems is straightforward and, since the equations are all initial-value, do not require large blocks of storage.

2 Characteristic Lengths for n-th Order Systems

Let us consider the system

$$u'(z) = A(z) u(z) + B(z) v(z), \qquad \text{(VIII.2.1a)}$$

$$-v'(z) = C(z) u(z) + D(z) v(z), \qquad \text{(VIII.2.1b)}$$

subject to

$$u(0) = 0, \quad u(x) = 0, \qquad \text{(VIII.2.2a, b)}$$

where u and v are n-vectors, the coefficients A, B, C, and D are $n \times n$ matrix functions of z and λ. Again, we assume that, for a fixed value of λ, the coefficients are such that there exist a countable number of values of x so that (VIII.2.1, VIII.2.2) have a nontrivial solution and that all initial-value problems for (VIII.2.1) have unique solutions.

The Riccati transformation for the above system is given by

$$\boldsymbol{u}\,(z) = R_1\,(z)\,\boldsymbol{v}\,(z), \tag{VIII.2.3}$$

where R_1 is an $n \times n$ matrix function. The more general transformation will be discussed in the next section. Although the foregoing transformation is valid, we simply do not have enough information available to derive the matrix differential equation satisfied by $R_1(z)$. In order to circumvent this problem, we must temporarily consider a more general problem. That is, we consider the system

$$U'\,(z) = A\,(z)\,U\,(z) + B\,(z)\,V\,(z), \tag{VIII.2.4a}$$

$$-\,V'\,(z) = C\,(z)\,U\,(z) + D\,(z)\,V\,(z), \tag{VIII.2.4b}$$

subject to

$$U\,(0) = 0, \quad V\,(0) = I, \tag{VIII.2.5a, b}$$

where U and V are $n \times n$ matrices, I is the $n \times n$ identity matrix, and A, B, C, and D are as given in (VIII.2.1). In terms of the matrix functions U and V, the transformation (VIII.2.3) becomes

$$U\,(z) = R_1\,(z)\,V\,(z), \tag{VIII.2.6}$$

where $R_1(z)$ is the same as in (VIII.2.3). The relationships between $\boldsymbol{u}$, $\boldsymbol{v}$ and U, V are given by

$$\boldsymbol{u}\,(z) = U\,(z)\,\boldsymbol{v}\,(0), \tag{VIII.2.7a}$$

$$\boldsymbol{v}\,(z) = V\,(z)\,\boldsymbol{v}\,(0). \tag{VIII.2.7b}$$

Now we can proceed to derive the matrix differential equation for R_1 in a fashion similar to that for the scalar functions in Chapter V. However, since we are dealing with matrices, the order of operation must be carefully observed. If we differentiate in (VIII.2.6), we get

$$U'\,(z) = R_1'\,(z)\,V\,(z) + R_1\,(z)\,V'\,(z). \tag{VIII.2.8}$$

Substituting (VIII.2.4) into (VIII.2.8), using (VIII.2.6) and simplifying, we get

$$\{R_1'(z) - B(z) - A(z) R_1(z) - R_1(z) D(z)$$
$$- R_1(z) C(z) R_1(z)\} V(z) = 0. \qquad \text{(VIII.2.9)}$$

Since $V(z)$ is nonsingular, at least in some neighborhood of the origin, it follows that the terms in braces must be zero. (It is now clear why we had to consider the more general matrix equations (VIII.2.4, VIIII.2.5).) Thus, we have

$$R_1'(z) = B(z) + A(z) R_1(z) + R_1(z) D(z)$$
$$+ R_1(z) C(z) R_1(z). \qquad \text{(VIII.2.10)}$$

The initial condition for (VIII.2.10) is

$$R_1(0) = 0, \qquad \text{(VIII.2.11)}$$

which follows by evaluating (VIII.2.6) at $z=0$.

In order to find the characteristic lengths of the problem (VIII.2.1, VIII.2.2), we evaluate (VIII.2.3) at $z = x$ and obtain

$$u(x) = 0 = R_1(x) v(x). \qquad \text{(VIII.2.12)}$$

For a nontrivial solution of (VIII.2.1, VIII.2.2) to exist, at least one component of $v(x)$ must be nonzero. Hence, it follows that, at $z = x$,

$$\det[R_1(x)] = 0. \qquad \text{(VIII.2.13)}$$

Also, by a reverse argument, it follows that the determinant of $R_1(z)$ is zero only at the characteristic lengths.

We cannot, however, simply integrate the matrix differential equation for $R_1(z)$ until the determinant vanishes. In the scalar case of Chapter V, we knew that the characteristic lengths for (V.2.1a–d) alternated with those of the problem (V.2.1a, b, V.2.9a, b), thus allowing for a simple switching procedure. This is not always the case for systems of equations due to the possible presence of several linearly independent solutions of (VIII.2.1, VIII.2.2) for certain values of the parameter λ. Hence, we shall temporarily assume that for each value of λ there is only one linearly independent solution of (VIII.2.1, VIII.2.2) and defer the discussion of multiple eigenfunctions to section 5.

The inverse transformation is defined by

$$v(z) = S_1(z) u(z), \qquad (VIII.2.14)$$

where $S_1(z)$ is an $n \times n$ matrix. In order to derive the differential equation for the $S_1(z)$ function, we must again use the techniques discussed earlier in this chapter. The equation is

$$- S_1'(z) = C(z) + S_1(z) A(z) + D(z) S_1(z)$$
$$+ S_1(z) B(z) S_1(z). \qquad (VIII.2.15)$$

On any common interval of definition the functions $R_1(z)$ and $S_1(z)$ are related by

$$S_1(z) = R_1^{-1}(z). \qquad (VIII.2.16)$$

The foregoing relation also serves as the initial condition for $S_1(z)$ after a switch has been performed.

It is clear that if we evaluate (VIII.2.14) at $z=z_1$, where $v(z_1)=0$, we obtain

$$0 = S_1(z_1) u(z_1). \qquad (VIII.2.17)$$

For this system of equations to have a nontrivial solution, it is necessary and sufficient that

$$\det [S_1(z_1)] = 0 \qquad (VIII.2.18)$$

Hence, we are also able to simultaneously compute the values of z_i, where (VIII.2.1), subject to the boundary conditions

$$u(0) = 0, \quad v(z_i) = 0, \qquad (VIII.2.19a, b)$$

has a nontrivial solution.

3   Modification of Boundary Conditions

If the boundary conditions are more complicated, the more general transformations must be introduced. Suppose we wish to find the characteristic lengths for (VIII.2.1), subject to the boundary conditions

$$\alpha_1 u(0) + \beta_1 v(0) + \gamma_1 u(x) + \delta_1 v(x) = 0, \qquad (VIII.3.1a)$$

$$\alpha_2 u(0) + \beta_2 v(0) + \gamma_2 u(x) + \delta_2 v(x) = 0, \qquad (VIII.3.1b)$$

where α_i, β_i, γ_i, and δ_i, $(i=1, 2)$ are $n \times n$ matrices which may contain λ.

The generalized Riccati and recovery transformations are then given by

$$u(z) = R_1(z) v(z) + R_2(z) u(0), \qquad \text{(VIII.3.2a)}$$

$$v(0) = Q_1(z) v(z) + Q_2(z) u(0), \qquad \text{(VIII.3.2b)}$$

where R_1, Q_1 are as before and R_2, Q_2 are $n \times n$ matrices which account for the fact that $u(0)$ will, in general, not be zero. The differential equations for R_1, R_2, Q_1, and Q_2 are derived in a fashion similar to those earlier in this chapter. They are

$$R_1'(z) = B(z) + A(z) R_1(z) + R_1(z) D(z)$$
$$\qquad + R_1(z) C(z) R_1(z), \qquad \text{(VIII.3.3a)}$$

$$R_2'(z) = [A(z) + R_1(z) C(z)] R_2(z), \qquad \text{(VIII.3.3b)}$$

$$Q_1'(z) = Q_1(z) [C(z) R_1(z) + D_1(z)], \qquad \text{(VIII.3.3c)}$$

$$Q_2'(z) = Q_1(z) [C(z) R_2(z)], \qquad \text{(VIII.3.3d)}$$

subject to the initial conditions

$$R_1(0) = 0, \quad R_2(0) = I, \quad Q_1(0) = I, \quad Q_2(0) = 0. \qquad \text{(VIII.3.4a-d)}$$

In order to calculate the characteristic lengths for (VIII.2.1), subject to (VIII.3.1), we view (VIII.3.1) and (VIII.3.2) as a system of $4n$ equations for the $4n$ unknowns $u(0)$, $v(0)$, $u(x)$, and $v(x)$. In matrix form we have

$$\begin{bmatrix} \alpha_1 & \beta_1 & \gamma_1 & \delta_1 \\ \alpha_2 & \beta_2 & \gamma_2 & \delta_2 \\ R_2(x) & 0 & -I & R_1(x) \\ Q_2(x) & -I & 0 & Q_1(x) \end{bmatrix} \begin{bmatrix} u(0) \\ v(0) \\ u(x) \\ v(x) \end{bmatrix} = \begin{bmatrix} 0 \\ 0 \\ 0 \\ 0 \end{bmatrix}. \qquad \text{(VIII.3.5)}$$

Thus, we seek the values of $z = x$ so that

$$\det(A) = \begin{vmatrix} \alpha_1 & \beta_1 & \gamma_1 & \delta_1 \\ \alpha_2 & \beta_2 & \gamma_2 & \delta_2 \\ R_2(x) & 0 & -I & R_1(x) \\ Q_2(x) & -I & 0 & Q_1(x) \end{vmatrix} = 0. \qquad \text{(VIII.3.6)}$$

Again, we must introduce the inverse Riccati and recovery transformations in order to avoid any singularities of the Riccati equation. Thus, at some point, z', we write

$$v(z) = S_1(z) u(z) + S_2(z) v(z'), \qquad \text{(VIII.3.7a)}$$

$$u(z') = T_1(z) u(z) + T_2(z) v(z'), \qquad \text{(VIII.3.7b)}$$

where S_1, S_2, T_1, and T_2 are $n \times n$ matrix functions which satisfy the initial-value problems

$$-S_1'(z) = C(z) + S_1(z) A(z) + D(z) S_1(z) + S_1(z) B(z) S_1(z),$$
$$\text{(VIII.3.8a)}$$

$$-S_2'(z) = [D(z) + S_1(z) B(z)] S_2(z), \qquad \text{(VIII.3.8b)}$$

$$S_1(z') = 0, \qquad \text{(VIII.3.8c)}$$

$$S_2(z') = I, \qquad \text{(VIII.3.8d)}$$

$$-T_1'(z) = T_1(z) [A(z) + B(z) S_1(z)], \qquad \text{(VIII.3.9a)}$$

$$-T_2'(z) = T_1(z) B(z) S_2(z), \qquad \text{(VIII.3.9b)}$$

$$T_1(z') - I, \qquad \text{(VIII.3.9c)}$$

$$T_2(z') = 0. \qquad \text{(VIII.3.9d)}$$

As in the scalar case we must rewrite (VIII.3.7) in terms of $u(0)$ and $v(0)$. We do this by evaluating (VIII.3.2) at $z = z'$ and solving for $u(z)'$ and $v(z')$. We find

$$v(z') = Q_1^{-1}(z') \{v(0) - Q_2(z') u(0)\}, \qquad \text{(VIII.3.10a)}$$

$$u(z') = R_1(z') Q_1^{-1}(z') v(0)$$
$$+ \{R_2(z') - R_1(z') Q_1^{-1}(z') Q_2(z')\} u(0). \qquad \text{(VIII.3.10b)}$$

Substitution of these values into (VIII.3.7) yields

$$v(z) = S_1(z) u(z) + S_2(z) \{A_1 v(0) + A_2 u(0)\}, \qquad \text{(VIII.3.11a)}$$

$$[B_2 - T_2(z) A_2] u(0) = T_1(z) u(z) + [T_2(z) A_1 - B_1] u(0), \qquad \text{(VIII.3.11b)}$$

where

$$A_1 = Q_1^{-1}(z'), \qquad A_2 = -Q_1^{-1}(z') Q_2(z'),$$

$$B_1 = R_1(z') Q_1^{-1}(z'), \qquad B_2 [R_2(z') - R_1(z') Q_1^{-1}(z') Q_2(z')].$$

The matrix $Q_1(z)$ is nonsingular, since it is the solution of a linear matrix differential equation with nonsingular initial conditions. The boundary conditions (VIII.3.1a, b) and the equations (VIII.3.11a, b) evaluated at $z=x$ form a system of $4n$ equations and $4n$ unknowns, $u(0)$, $v(0)$, $u(x)$, and $v(x)$. In matrix form these equations become

$$\begin{bmatrix} \alpha_1 & \beta_1 & \gamma_1 & \delta_1 \\ \alpha_2 & \beta_2 & \gamma_2 & \delta_2 \\ S_2(x)A_2 & S_2(x)A_1 & S_1(x) & -I \\ T_2(x)A_2 - B_2 & T_2(x)A_1 - B_1 & T_1(x) & 0 \end{bmatrix} \begin{bmatrix} u(0) \\ v(0) \\ u(x) \\ v(x) \end{bmatrix} = \begin{bmatrix} 0 \\ 0 \\ 0 \\ 0 \end{bmatrix}.$$

$$\text{(VIII.3.12)}$$

Thus, the characteristic lengths are the values of x so that the determinant of the matrix of coefficients in (VIII.3.12) is zero. We may summarize our results as a theorem.

Theorem 1: If the coefficients A, B, C, and D are real and are such that the solutions of (VIII.2.1a, b), subject to the boundary conditions (VIII.3.1a, b), have a countable number of points, x_i, for which the solutions are nontrivial, these points correspond to the points where the determinant of the matrix of coefficients of (VIII.3.5) or (VIII.3.12) is zero.

4   An Analytical Example

An equation which frequently appears in various studies of structural mechanics is

$$y^{(4)} = k^4 y. \tag{VIII.4.1a}$$

A typical set of boundary conditions might be

$$y(0) = 0, \tag{VIII.4.1b}$$

$$y''(0) = 0, \tag{VIII.4.1c}$$

$$y(x) = 0, \tag{VIII.4.1d}$$

$$y''(x) = 0. \tag{VIII.4.1e}$$

The general solution of (VIII.4.1a) is given by

$$y(z) = A \sin kz + B \sinh kz + C \cos kz + D \cosh kz. \tag{VIII.4.2}$$

With the particular boundary conditions given in (VIII.4.1b–e), the system will have a nontrivial solution if and only if $x \in \{x : x = n\pi/k, n = 1, 2, \ldots\}$. The eigenfunctions are then

$$y(z) = A \sin kz, \qquad \text{(VIII.4.3)}$$

where A is arbitrary.

In order to apply the method of invariant imbedding to the above problem we first rewrite (VIII.4.1a–e) as a system of four first-order equations. Although there are many ways to reduce an nth-order equation to a system of n first order equations, we wish to choose a method which may reduce the complexity of the Riccati and recovery transformations.

For example, if we let $u_1 = y$, $u_2 = y''$, $v_1 = y'$, $v_2 = y'''$, then our system becomes

$$\frac{d}{dz}\begin{pmatrix} u_1 \\ u_2 \end{pmatrix} = \begin{pmatrix} 1 & 0 \\ 0 & 1 \end{pmatrix}\begin{pmatrix} v_1 \\ v_2 \end{pmatrix}, \qquad \text{(VIII.4.4a)}$$

$$-\frac{d}{dz}\begin{pmatrix} v_1 \\ v_2 \end{pmatrix} = \begin{pmatrix} 0 & -1 \\ -k^4 & 0 \end{pmatrix}\begin{pmatrix} u_1 \\ u_2 \end{pmatrix}, \qquad \text{(VIII.4.4b)}$$

$$\begin{pmatrix} u_1(0) \\ u_2(0) \end{pmatrix} = \begin{pmatrix} 0 \\ 0 \end{pmatrix} \qquad \text{(VIII.4.4c)}$$

$$\begin{pmatrix} u_1(x) \\ u_2(x) \end{pmatrix} = \begin{pmatrix} 0 \\ 0 \end{pmatrix}. \qquad \text{(VIII.4.4d)}$$

The Riccati and recovery transformations may then be written as

$$\begin{pmatrix} u_1(z) \\ u_2(z) \end{pmatrix} = \begin{pmatrix} r_{11}(z) & r_{12}(z) \\ r_{21}(z) & r_{22}(z) \end{pmatrix}\begin{pmatrix} v_1(z) \\ v_2(z) \end{pmatrix} \qquad \text{(VIII.4.5a)}$$

$$\begin{pmatrix} v_1(0) \\ v_2(0) \end{pmatrix} = \begin{pmatrix} q_{11}(z) & q_{12}(z) \\ q_{21}(z) & q_{22}(z) \end{pmatrix}\begin{pmatrix} v_1(z) \\ v_2(z) \end{pmatrix} \qquad \text{(VIII.4.5b)}$$

The recovery transformation, (VIII.4.5b), is needed only if we wish to calculate the eigenfunctions of the problem.

However, if we use the more obvious substitution $u_1 = y$, $u_2 = y'$, $v_1 = y''$, and $v_2 = y'''$, then our system becomes

$$\frac{d}{dz}\begin{pmatrix} u_1(z) \\ u_2(z) \end{pmatrix} = \begin{pmatrix} 0 & 1 \\ 0 & 0 \end{pmatrix}\begin{pmatrix} u_1(z) \\ u_2(z) \end{pmatrix} + \begin{pmatrix} 0 & 0 \\ 1 & 0 \end{pmatrix}\begin{pmatrix} v_1(z) \\ v_2(z) \end{pmatrix}, \qquad \text{(VIII.4.6a)}$$

$$-\frac{d}{dz}\begin{pmatrix} v_1(z) \\ v_2(z) \end{pmatrix} = \begin{pmatrix} 0 & 0 \\ -k^4 & 0 \end{pmatrix}\begin{pmatrix} u_1(z) \\ u_2(z) \end{pmatrix} + \begin{pmatrix} 0 & -1 \\ 1 & 0 \end{pmatrix}\begin{pmatrix} v_1(z) \\ v_2(z) \end{pmatrix}, \qquad \text{(VIII.4.6b)}$$

$$\begin{pmatrix} 1 & 0 \\ 0 & 0 \end{pmatrix} \begin{pmatrix} u_1(0) \\ u_2(0) \end{pmatrix} + \begin{pmatrix} 0 & 0 \\ 1 & 0 \end{pmatrix} \begin{pmatrix} v_1(0) \\ v_2(0) \end{pmatrix} = \begin{pmatrix} 0 \\ 0 \end{pmatrix}, \qquad \text{(VIII.4.6c)}$$

$$\begin{pmatrix} 1 & 0 \\ 0 & 0 \end{pmatrix} \begin{pmatrix} u_1(x) \\ u_2(x) \end{pmatrix} + \begin{pmatrix} 0 & 0 \\ 1 & 0 \end{pmatrix} \begin{pmatrix} v_1(x) \\ v_2(x) \end{pmatrix} = \begin{pmatrix} 0 \\ 0 \end{pmatrix}. \qquad \text{(VIII.4.6d)}$$

Notice that the matrices A and D now have nonzero entries and the boundary conditions are more complicated. Hence, we must either use a further substitution as discussed in section V.5 or use the Riccati and recovery transformations in the form

$$\begin{pmatrix} u_1(z) \\ u_2(z) \end{pmatrix} = \begin{pmatrix} r_{11}^{(1)}(z) & r_{12}^{(1)}(z) \\ r_{21}^{(1)}(z) & r_{22}^{(1)}(z) \end{pmatrix} \begin{pmatrix} v_1(z) \\ v_2(z) \end{pmatrix} + \begin{pmatrix} r_{11}^{(2)}(z) & r_{12}^{(2)}(z) \\ r_{21}^{(2)}(z) & r_{22}^{(2)}(z) \end{pmatrix} \begin{pmatrix} u_1(0) \\ u_2(0) \end{pmatrix}$$
$$\text{(VIII.4.7a)}$$

$$\begin{pmatrix} v_1(0) \\ v_2(0) \end{pmatrix} = \begin{pmatrix} q_{11}^{(1)}(z) & q_{12}^{(1)}(z) \\ q_{21}^{(1)}(z) & q_{22}^{(1)}(z) \end{pmatrix} \begin{pmatrix} v_1(z) \\ v_2(z) \end{pmatrix} + \begin{pmatrix} q_{11}^{(2)}(z) & q_{12}^{(2)}(z) \\ q_{21}^{(2)}(z) & q_{22}^{(2)}(z) \end{pmatrix} \begin{pmatrix} u_1(0) \\ u_2(0) \end{pmatrix}.$$
$$\text{(VIII.4.7b)}$$

Thus we prefer the formulation given by (VIII.4.4, VIII.4.5).

The matrix differential equations satisfied by the r functions in (VIII.4.5a) are given by

$$r_{11}' = 1 - r_{11}r_{21} - k^4 r_{11}r_{12}, \qquad r_{11}(0) = 0, \qquad \text{(VIII.4.8a)}$$

$$r_{12}' = -r_{11}r_{22} - k^4 r_{12}^2, \qquad r_{12}(0) = 0, \qquad \text{(VIII.4.8b)}$$

$$r_{21}' = -r_{21}^2 - k^4 r_{11}r_{22}, \qquad r_{21}(0) = 0, \qquad \text{(VIII.4.8c)}$$

$$r_{22}' = 1 - r_{22}r_{21} - k^4 r_{12}r_{22}, \qquad r_{22}(0) = 0. \qquad \text{(VIII.4.8d)}$$

The solutions of these equations are

$$r_{11}(z) = \frac{1}{2k} (\tanh kz + \tan kz), \qquad \text{(VIII.4.9a)}$$

$$r_{12}(z) = \frac{1}{2k^3} (\tanh kz - \tan kz), \qquad \text{(VIII.4.9b)}$$

$$r_{21}(z) = \frac{k}{2} (\tanh kz - \tan kz), \qquad \text{(VIII.4.9c)}$$

$$r_{22}(z) = \frac{1}{2k} (\tanh kz + \tan kz). \qquad \text{(VIII.4.9d)}$$

In this particular example the criterion given in (VIII.3.5) for finding the characteristic lengths reduces to finding the values of z where

$$0 = |R_1| = \begin{vmatrix} r_{11} & r_{12} \\ r_{21} & r_{22} \end{vmatrix} = \begin{vmatrix} \dfrac{1}{2k}(\tanh kz + \tan kz) & \dfrac{1}{2k^3}(\tanh kz - \tan kz) \\ \dfrac{k}{2}(\tanh kz - \tan kz) & \dfrac{1}{2k}(\tanh kz + \tan kz) \end{vmatrix}$$

$$= \frac{1}{k^2} \tan kz \tanh kz.$$

These values of $z = x$ are easily seen to be multiples of π/k. This agrees with the classical result given earlier.

The inverse transformations must, of course, be used to integrate through the various singularities.

5 Multiple Eigenfunctions

In the case of single second order equations there normally exists only one eigenfunction, apart from constant multiples, for each eigenvalue. In fact, the only time that multiple eigenfunctions can occur for single second-order equations is when the boundary conditions are nonseparated, such as when periodic. However, for systems the problem is considerably more complicated. There are no simple criteria for determining *a priori* when an eigenvalue of a general nth-order equation with prescribed boundary conditions will have several linearly independent eigenfunctions. We shall see, however, that the presence of multiple eigenfunctions causes no particular problem in our calculational procedure.

The procedure described in the previous section must be slightly modified, since the singularities and zeros of the determinant of the matrix of coefficients will not necessarily separate one another. This simply implies that the switching process should not be performed. The presence of multiple eigenfunctions is easily detected, since the multiplicity of the roots at x, where the determinant is zero, indicates the multiplicity of the linearly independent eigenfunctions.

In order to illustrate the behavior of the determinant of the matrix of coefficients when a degenerate eigenvalue is present, we shall consider three examples. First, let us consider the problem discussed in the previous section

$$y^{iv} = k^4 y, \tag{VIII.5.1}$$

$$y(0) = y''(0) = 0 = y(x) = y''(x). \tag{VIII.5.2}$$

This particular problem has only simple eigenvalues. For a given value of k, the search for the values of x so that (VIII.5.1, VIII.5.2) has a nontrivial solution amounts to finding a value of x so that

$$0 = |A| = |R_1| = \frac{1}{k^2} \tan kx \tanh kx. \tag{VIII.5.3}$$

In Figure 8-1 we have plotted the value of R_1 versus the variable x, for $k=1$. The values of $z=x$, where $|R_1|=0$ are easily seen to be multiples of π.

The second example,

$$y^{iv} + 4y = -\lambda y'' \tag{VIII.5.4}$$

$$y(0) = y''(0) = 0 = y(x) = y''(x). \tag{VIII.5.5}$$

has the two eigenfunctions $\sin x$ and $\sin 2x$ when $\lambda = 5$. In this case the deter-

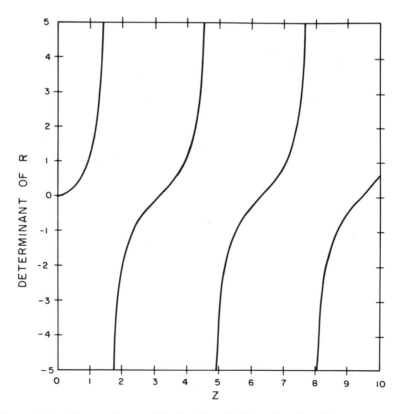

Figure 8-1.   Determinant of R_1 for Simple Eigenfunction Case

minant of $|R_1|$ is

$$|R_1| = \tfrac{1}{2} \tan x \tan 2x = \frac{\sin^2 x}{\cos 2x}. \qquad \text{(VIII.5.6)}$$

The above function has a double root at multiples of π indicating the existence of two linearly independent eigenfunctions. These results are illustrated in Figure 8-2.

The last example to be considered is given by

$$-y^{iv} - 49y'' = \lambda(14y^{iv} + 36y), \qquad \text{(VIII.5.7)}$$

$$y(0) = y''(0) = y^{iv}(0) = 0 = y(x) = y''(x) = y^{iv}(x). \qquad \text{(VIII.5.8)}$$

For $\lambda = 1$, the problem has the three eigenfunctions $\sin kx$, $k = 1, 2, 3$. Here

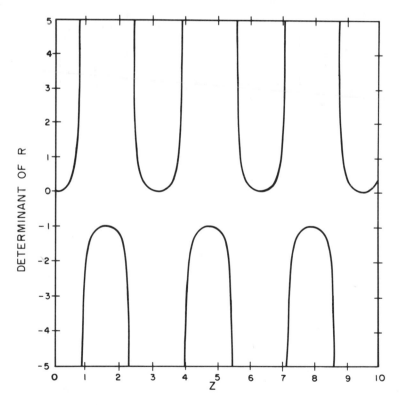

Figure 8-2. Determinant of R_1 for Double Eigenfunction Case

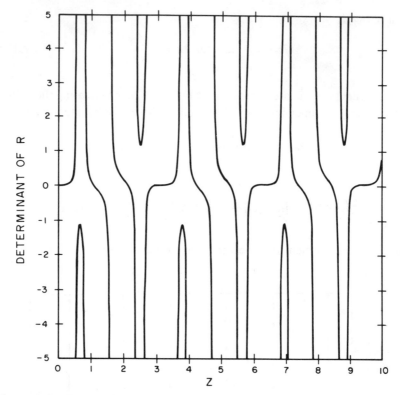

Figure 8-3. Determinant of R_1 for Triple Eigenfunction Case

the determinant of R_1 is given by

$$|R_1| = \tfrac{1}{6} \tan x \tan 2x \tan 3x = \frac{\sin^3 x \,(3 - 4 \sin^2 x)}{3 \cos x \cos 2x \,(4 \cos^2 x - 3)}.$$ (VIII.5.9)

The determinant of R_1 is plotted in Figure 8-3. Notice that at multiples of π, the determinant of R_1 has a triple root indicating the existence of three linearly independent eigenfunctions for $\lambda = 1$.

6 Inhomogeneous Equations

Consider the problem

$$\boldsymbol{u}'(z) = A(z)\,\boldsymbol{u}(z) + B(z)\,\boldsymbol{v}(z) + \boldsymbol{e}(z),$$ (VIII.6.1a)

$$-v'(z) = C(z)\,u(z) + D(z)\,v(z) + f(z), \qquad \text{(VIII.6.1b)}$$

$$u(0) = 0, \qquad v(x) = a. \qquad \text{(VIII.6.2a, b)}$$

Here u and v are an m-vector and an n-vector, respectively, and the remaining quantities are given matrices or vectors of the appropriate dimensions. We shall consider only values of x lying in the range $0 \leqslant x \leqslant X$, where $X > 0$ is sufficiently small that the problem (VIII.6.1) has a unique solution for all such x. Problems involving $x > X$ could be considered in the manner of section 3 of Chapter VI.

Let the $m \times n$ matrix $U(z)$ and $n \times n$ matrix $V(z)$ be defined by the homogeneous initial value problem

$$U'(z) = A(z)\,U(z) + B(z)\,V(z), \qquad \text{(VIII.6.3a)}$$

$$-V'(z) = C(z)\,U(z) + D(z)\,V(z), \qquad \text{(VIII.6.3b)}$$

$$U(0) = 0, \qquad V(0) = I. \qquad \text{(VIII.6.4a, b)}$$

The matrix $V(z)$ is nonsingular for all z so that $z \leqslant X$. Let $\hat{u}(z)$, $\hat{v}(z)$ be the solution of (VIII.6.1) with the initial conditions $\hat{u}(0) = 0$, $\hat{v}(0) = 0$. Then the principle of superposition gives the solution of (VIII.6.1, VIII.6.2) as

$$u(z) = U(z)\,V^{-1}(x)\,[a - \hat{v}(z)] + \hat{u}(z), \qquad \text{(VIII.6.5a)}$$

$$v(z) = V(z)\,V^{-1}(x)\,[a - \hat{v}(z)] + \hat{v}(z). \qquad \text{(VIII.6.5b)}$$

Let us introduce the Riccati and recovery transformations

$$u(z) = R(z)\,v(z) + r(z), \qquad \text{(VIII.6.6a)}$$

$$v(0) = Q(z)\,v(z) + q(z), \qquad \text{(VIII.6.6b)}$$

where $R(z)$ is an $m \times n$ matrix, r is an m-vector, Q is an $n \times n$ matrix, and q is an n-vector. Using the techniques of section 2, we find that the above functions satisfy the differential equations

$$R'(z) = B(z) + A(z)\,R(z) + R(z)\,D(z) + R(z)\,C(z)\,R(z),$$
$$\text{(VIII.6.7a)}$$

$$r'(z) = [A(z) + R(z)\,C(z)]\,r(z) + R(z)\,f(z) + e(z), \qquad \text{(VIII.6.7b)}$$

$$Q'(z) = Q(z)\,[C(z)\,R_1(z) + D(z)], \qquad \text{(VIII.6.7c)}$$

$$q'(z) = Q(z)\,[f(z) + C(z)\,r(z)], \qquad \text{(VIII.6.7d)}$$

$$R(0) = 0, \qquad r(0) = q(0) = 0 \qquad \text{(VIII.6.8a–d)}$$

By evaluating (VIII.6.6b) at $z = x$ and then solving for $u(z)$ and $v(z)$, we obtain

$$v(z) = Q^{-1}(z) [Q(x) a + q(x) - q(z)], \qquad \text{(VIII.6.9a)}$$

$$u(z) = R(z) v(z) + r(z). \qquad \text{(VIII.6.9b)}$$

We can now give the relationships among the R and Q functions and the super position functions

$$R_1(z) = U(z) V^{-1}(z), \qquad Q_1(z) = V^{-1}(z), \qquad \text{(VIII.6.10a, b)}$$

$$r(z) = \hat{u}(z) - U(z) V^{-1}(z) \hat{v}(z), \qquad q(z) = - V^{-1}(z) \hat{v}(z). \qquad \text{(VIII.6.10c, d)}$$

Up to this point the endpoint x has been regarded as fixed. We now regard x as a variable and introduce the definitions

$$J(z, x) = U(z) V^{-1}(x), \qquad K(z, x) = V(z) V^{-1}(x), \qquad \text{(VIII.6.11a, b)}$$

$$m(z, x) = \hat{u}(z) = U(z) V^{-1}(x) \hat{v}(x), \qquad \text{(VIII.6.11c)}$$

$$n(z, x) = \hat{v}(z) - V(z) V^{-1}(x) \hat{v}(x). \qquad \text{(VIII.6.11d)}$$

As functions of z for arbitrary fixed x, J and K obviously satisfy the homogeneous matrix differential system (VIII.6.3) and the boundary conditions

$$J(0, x) = 0, \qquad K(x, x) = I. \qquad \text{(VIII.6.12a, b)}$$

Similarly, as functions of z for any fixed x, the vectors m and n satisfy the inhomogeneous system (VIII.6.1) and the boundary conditions

$$m(0, x) = 0, \qquad n(x, x) = 0. \qquad \text{(VIII.6.13a, b)}$$

It is easy to see that the solution of (VIII.6.1, VIII.6.2) is given by

$$u(z) = m(z, x) + J(z, x) a, \qquad \text{(VIII.6.14a)}$$

$$v(z) = n(z, x) + K(z, x) a \qquad \text{(VIII.6.14b)}$$

in terms of the functions defined in the previous paragraph. Let us now denote partial differentiation by subscripts in the standard manner. Then J, K, m, and n satisfy the following partial differential system:

$$J_2(z, x) = J(z, x) [C(x) R(x) + D(x)], \qquad \text{(VIII.6.15a)}$$

$$K_2(z, x) = K(z, x) [C(x) R(x) + D(x)], \qquad \text{(VIII.6.15b)}$$

$$m_2(z, x) = J(z, x) [C(x) r(x) + f(x)], \qquad \text{(VIII.6.15c)}$$

$$n_2(z, x) = K(z, x) [C(x) r(x) + f(x)]. \qquad \text{(VIII.6.15d)}$$

The initial conditions are

$$J(z, z) = R(z), \quad K(z, z) = I, \qquad \text{(VIII.6.16a, b)}$$

$$m(z, z) = r(z), \quad n(z, z) = 0. \qquad \text{(VIII.6.16c, d)}$$

These follow easily from (VIII.6.10) and (VIII.6.11) (see exercise 4).

Now, for fixed z and for x, $0 \leqslant z \leqslant x$, we can obtain $J(z, x)$, $K(z, x)$, $m(z, x)$ and $n(z, x)$ and, hence, by (VIII.6.14) the solution of (VIII.6.1, VIII.6.2) as follows. First, the initial-value problem consisting of (VIII.6.7a–c), with initial conditions from (VIII.6.8), is numerically integrated from 0 to z. Then the system (VIII.6.15), regarded as an ordinary differential system in x, with initial values at $x = z$ obtained from (VIII.6.16), is adjoined to the system (VIII.6.7a–c), and this larger system is numerically integrated from $x = z$ to the desired interval length.

From (VIII.6.10) and (VIII.6.11) it is seen that following relationships express the fundamental quantities of either of the methods in terms of those appearing in the other method:

$$R(z) = J(z, z), \quad Q(z) = K(0, z), \qquad \text{(VIII.6.17a, b)}$$

$$r(z) = m(z, z), \quad q(z) = n(0, z), \qquad \text{(VIII.6.17c, d)}$$

$$J(z, x) = R(z) Q^{-1}(z) Q(x), \quad K(z, x) = Q^{-1}(z) Q(x), \qquad \text{(VIII.6.18a, b)}$$

$$m(z, x) = r(z) + R(z) Q^{-1}(z) [q(x) - q(z)], \qquad \text{(VIII.6.18c)}$$

$$n(z, x) = Q^{-1}(z) [q(x) - q(z)]. \qquad \text{(VIII.6.18d)}$$

7 Numerical Examples

We shall consider several examples of systems of various orders. Extreme accuracy was not attempted, since the studies were primarily of a feasibility nature. All of the examples were generated with a single program, utilizing a standard fourth-order Runge-Kutta integration scheme with fixed step size on a CDC-6600.

Example 1

Alspaugh, Kagiwada, and Kalaba recently considered the calculation of the eigenvalues for buckling of columns. The problem considered was a column of stiffness, EI, and length, L_1, loaded by an axial compressive force of magnitude, P, and supported laterally by an elastic foundation of spring constant, k. The governing equation of equilibrium is given by

$$\frac{d^2}{dz^2}\left(EI\frac{d^2w}{dz^2}\right) + P\frac{d^2}{dz^2} + kw = 0, \qquad \text{(VIII.7.1)}$$

where w is the transverse displacement of the neutral axis. The various types of boundary conditions to be imposed are

$$\text{simple support} \quad w = 0, \quad \frac{d^2w}{dz^2} = 0,$$

$$\text{clamped support} \quad w = 0, \quad \frac{dw}{dz} = 0,$$

$$\text{free} \quad \frac{d^2w}{dz^2} = 0, \quad \frac{d}{dz}\left(EI\frac{d^2w}{dz^2}\right) + P\frac{dw}{dz} = 0.$$

In section III of the paper by Alspaugh, Kagiwada, and Kalaba, the following is stated; "It is characteristic of invariant imbedding that the equations must be derived for each set of boundary conditions used." Although the statement is true for the classical form, it is not true of the form of the imbedding used in this book. In addition, the classical form of the imbedding will occasionally give infinite initial conditions for the Riccati equations. However, the use of a judicous substitution or the introduction of the more general transformations given by (VIII.3.2) eliminates this problem. This requires, of course, the integration of more differential equations, which also may be used in determining the eigenfunctions.

Let us consider a uniform column which is simply supported at both ends. Equation (VIII.7.1) then becomes

$$w^{iv} + \frac{P}{EI}w'' + \frac{k}{EI}w = 0, \qquad \text{(VIII.7.2a)}$$

subject to

$$w(0) = 0, \quad w''(0) = 0, \quad w(L_1) = 0, \quad w''(L_1) = 0. \quad \text{(VIII.7.2b)}$$

in order to write (VIII.7.2) as a system of first order equations, we let

$$\begin{aligned} u_1(z) &= w(z), \\ u_2(z) &= w''(z), \\ v_1(z) &= w'(z), \\ v_2(z) &= w'''(z). \end{aligned} \quad \text{(VIII.7.3)}$$

Then (VIII.7.2) becomes

$$\begin{pmatrix} u_1' \\ u_2' \end{pmatrix} = \begin{pmatrix} 1 & 0 \\ 0 & 1 \end{pmatrix} \begin{pmatrix} v_1 \\ v_2 \end{pmatrix}$$

$$-\begin{pmatrix} v_1' \\ v_2' \end{pmatrix} = \begin{pmatrix} 0 & -1 \\ \dfrac{k}{EI} & \dfrac{P}{EI} \end{pmatrix} \begin{pmatrix} u_1 \\ u_2 \end{pmatrix} \quad \text{(VIII.7.4a)}$$

subject to

$$\begin{pmatrix} u_1(0) \\ u_2(0) \end{pmatrix} = \begin{pmatrix} 0 \\ 0 \end{pmatrix}, \begin{pmatrix} u_1(L_1) \\ u_2(L_1) \end{pmatrix} = \begin{pmatrix} 0 \\ 0 \end{pmatrix}. \quad \text{(VIII.7.4b)}$$

Since the vector $u(0)$ is zero, we do not need the more general transformations, and the analysis presented in section 3 is applicable. If we let $P=0$ and $k/EI = -k^4$, then the solution of (VIII.7.4) is given by (VIII.4.3).

Suppose the boundary conditions are free at $z=0$ and clamped at $z=L_1$. An appropriate set of substitutions is

$$\begin{aligned} u_1(z) &= EIw''(z), \\ u_2(z) &= EIw'''(z) + Pw'(z), \\ v_1(z) &= w(z), \\ v_2(z) &= w'(z), \end{aligned} \quad \text{(VIII.7.5)}$$

and we have

$$\begin{pmatrix} u_1' \\ u_2' \end{pmatrix} = \begin{pmatrix} 0 & 1 \\ 0 & 0 \end{pmatrix} \begin{pmatrix} u_1 \\ u_2 \end{pmatrix} + \begin{pmatrix} 0 & -P \\ -K & 0 \end{pmatrix} \begin{pmatrix} v_1 \\ v_2 \end{pmatrix}$$

$$-\begin{pmatrix} v_1' \\ v_2' \end{pmatrix} = \begin{pmatrix} 0 & 0 \\ -1/EI & 0 \end{pmatrix} \begin{pmatrix} u_1 \\ u_2 \end{pmatrix} + \begin{pmatrix} 0 & -1 \\ 0 & 0 \end{pmatrix} \begin{pmatrix} v_2 \\ v_2 \end{pmatrix}, \quad \text{(VIII.7.6a)}$$

$$\begin{pmatrix} u_1(0) \\ u_2(0) \end{pmatrix} = \begin{pmatrix} 0 \\ 0 \end{pmatrix}, \begin{pmatrix} v_1(L_1) \\ v_2(L_1) \end{pmatrix} = \begin{pmatrix} 0 \\ 0 \end{pmatrix}. \quad \text{(VIII.7.6b)}$$

Again we do not need the more general transformations. Calculations were performed for several values of the parameters, P, E, I, and k. In all cases the first five characteristic lengths were computed to within six significant digits of the analytical results.

Example 2

We now wish to discuss the calculation of the characteristic lengths for the problem

$$y^{(100)} = ky, \tag{VIII.7.7a}$$

$$y(0) = 0, \quad y(x) = 0, \tag{VIII.7.7b}$$

$$y^{(2i)}(0) = 0, \quad y^{(2i)}(x) = 0, \quad i = 1, 2, ..., 50.$$

This example was chosen to demonstrate the feasibility of the method for high-order systems which occur in the study of multigroup diffusion equations of neutron physics. For $k = 1$, the characteristic lengths are simply $x_n = n\pi$.

	True Value	*Calculated Value*
$x_1 =$	3.141593	3.141591
$x_2 =$	6.283185	6.283177
$x_3 =$	9.424780	9.424772

Example 3

The following two problems arise in the study of elastic beams:

$$y^{iv} + 4y = -\lambda y'' \tag{VIII.7.8a}$$

$$y(0) = 0, \quad y(x) = 0, \tag{VIII.7.8b}$$

$$y''(0) = 0, \quad y''(x) = 0,$$

and

$$-y^{vi} - 49y'' = \lambda(14y^{iv} + 36y) \tag{VIII.7.9a}$$

$$y(0) = 0, \quad y(x) = 0,$$

$$y''(0) = 0, \quad y''(x) = 0, \tag{VIII.7.9b}$$

$$y^{iv}(0) = 0, \quad y^{iv}(x) = 0.$$

Both of these problems have, for certain values of λ, multiple eigenfunctions. When $\lambda = 5$, the first problem has the two eigenfunctions $y_1 = \sin z$ and $y_2 = \sin 2z$; whereas for $\lambda = 1$, the second problem has the three eigenfunctions

$y_1 = \sin z$, $y_2 = \sin 2z$, and $y_3 = \sin 3z$. Again we emphasize, as we did in section 5, that the presence of multiple eigenfunctions causes no particular difficulty. In fact, we obtained six significant digits in the calculation of the first three characteristic lengths for both problems.

Example 4

We shall now demonstrate how the method can be used to compute the eigenvalues or characteristic lengths for certain integrodifferential equations. Consider the equation

$$(\text{sgn } s)\frac{\partial n}{\partial z}(z, s) + a(s)n(z, s) = \lambda k(s)\int_{-1}^{1} n(z, s)\,ds', \quad 0 \leqslant z \leqslant x, \tag{VIII.7.10a}$$

subject to the conditions

$$n(0, s) = 0, \qquad 0 < s < 1, \tag{VIII.7.10b}$$

$$n(x, s) = 0, \qquad -1 < s < 0, \tag{VIII.7.10c}$$

where $a(s)$ and $k(s)$ are real piecewise continuous functions on $|s| \leqslant 1$. For fixed λ, we wish to compute the interval lengths x so that (VIII.7.10) has a nontrivial solution. Equations of the above form arise in the study of particle transport in a slab.

In order to apply our techniques to this type of problem, we make the following substitutions

$$u(z, s) = n(z, s), \qquad \text{when } s > 0, \tag{VIII.7.11a}$$

$$v(z, s) = n(z, s), \qquad \text{when } s < 0. \tag{VIII.7.11b}$$

Then (VIII.7.10) can be written as

$$\frac{\partial u}{\partial v}(z, s) + a(s)u(z, s) = \lambda k(s)\left\{\int_{0}^{1} u(z, s')\,ds' + \int_{-1}^{0} v(z, s')\,ds'\right\} \tag{VIII.7.12a}$$

$$\frac{\partial v}{\partial z}(z, s) + a(s)v(z, s) = \lambda k(s)\left\{\int_{0}^{1} u(z, s')\,ds' + \int_{-1}^{0} v(z, s')\,ds'\right\} \tag{VIII.7.12b}$$

$$u(0, s) = 0, \tag{VIII.7.12c}$$

$$v(x, s) = 0. \tag{VIII.7.12d}$$

The integrals are replaced with some type of numerical quadrature, such as Gaussian quadrature, and then (VIII.7.12) becomes a system of ordinary differential equation of the form

$$u' = Au + Bv, \qquad (VIII.7.13a)$$

$$- v' = Cu + Dv, \qquad (VIII.7.13b)$$

$$u(0) = 0, \qquad (VIII.7.13c)$$

$$v(x) = 0. \qquad (VIII.7.13d)$$

We shall consider two examples of this class of problem. The first example has been studied by Wing and Allen. Let

$$a(s) = |s|,$$

$$k(s) = e^{-5|s|}.$$

The results of our computation are compared in Table VIII.7-1 with those obtained by Allen and Wing.

The second example of the integrodifferential equation type is the Boltzmann equation for particle transport in a slab, where

$$a(s) = \frac{1}{|s|},$$

$$k(s) = \frac{1}{2|s|}.$$

TABLE VIII.7-1.   *Comparison of Several Techniques for the Computation of Characteristic Lengths for a Boltzmann-Type Equation*

	Scott	Wing	Allen
λ	x_1	x_1	x_1
40	0.12687	0.13672	0.1318
30	0.16962	NR*	0.1766
20	0.25581	0.26562	0.2680
10	0.52000	0.52930	0.5452
5	1.10744	1.08398	1.1292
2	2.9617	2.97266	NR*

* NR means no results were given for this case.

We shall compare our results with two separate techniques. Wing has performed extensive calculations for upper and lower bounds on the eigenvalues of this equation. We present two of his calculations

x	$\underline{\mu}_1$	$\bar{\mu}_1$
0.2	0.261	0.262
2.0	0.783	0.785,

where $\underline{\mu}_1$ and $\bar{\mu}_1$ are, respectively, lower and upper bounds on the first eigenvalue.

Our calculations are as follows:

$$\mu=0.2610 \qquad x_1=0.199937$$
$$\mu=0.2615 \qquad x_1=0.200533$$
$$\mu=0.2620 \qquad x_1=0.201131,$$

and

$$\mu=0.783 \qquad x_1=1.999778$$
$$\mu=0.784 \qquad x_1=2.00927$$
$$\mu=0.785 \qquad x_1=2.01883,$$

where $\mu=1/\lambda$ in (VIII.7.10a).

Our calculations are obviously consistent with those of Wing. The last comparison is made with the results of Carlson and Bell on the calculation of the critical dimensions of a particle-multiplying medium in slab geometry. We emphasize that the criticality type of computations are ideally suited to the invariant imbedding approach.

λ	*Scott*	*Carlson and Bell*
2.0	0.62205	0.6216

The result of Carlson and Bell was obtained by using a very sophisticated variational approach.

8   Exercises

1. Show that $J(z, x)$ and $K(z, x)$, as defined by (VIII.6.11), satisfy (VIII.6.3) as a function of z and fixed x.

2. Show that $m(z, x)$ and $n(z, x)$, as defined by (VIII.6.11), satisfy (VIII.6.1) as a function of z and fixed x.

3. Show that $u(z, x)$ and $v(z, x)$, as defined by (VIII.6.14), satisfy (VIII.6.1) and (VIII.6.2).

4. Verify (VIII.6.16) by using (VIII.6.10) and (VIII.6.11).

5. Derive (VIII.2.15).

6. Verify that (VIII.4.9a–d) are solutions of (VIII.4.8a–d).

7. Verify that (VIII.6.9a–d) are solutions of (VIII.6.1a–d).

9 Bibliographical Discussion

Sections 1–5

The material of this section first appeared in

M. R. Scott, *An Initial Value Method for the Eigenvalue Problem for Ordinary Differential Equations*, Sandia Laboratories, SC-RR-71-0791, August 1972.

A matrix Prüfer transformation formulation can be found in

J. M. Calvert and W. D. Davison, "Oscillation Theory and Computational Procedures for Matrix Sturm-Liouville Eigenvalue Problems, with an Application to the Hydrogen Ion," *J. Phys. Series A* **2** (1969), 278–292.

Section 6

The connection between superposition and the two basic versions of invariant imbedding is presented. See also

P. Nelson, Jr. and M. R. Scott, "The Relationship Between Two Variants of Invariant Imbedding," *J. Math. Anal. Appl.* **37** (1972), 501–505.

Section 7

Example 1

D. W. Alspaugh, H. H. Kagiwada, and R. E. Kalaba, *Application of Invariant Imbedding to the Eigenvalue Problems for Buckling of Columns*, The Rand Corp. RM-5954-PR. 1969.

Example 3

L. Collatz, *Eigenwertaufgaben mit Technischen Anwendungen*, Akademische Verlagsgesellschaft, Leipzig, 1963.

Example 4

R. C. Allen, Jr., "Functional Relationships for Fredholm Integral Equations Arising From Pseudo-Transport Problems," *J. Math. Anal. Appl.* **30** (1971), 48–78.

B. G. Carlson and G. I. Bell, "Solution of the Transport Equation by the S_n Method," *Proc. Intern. Conf. Peaceful Uses of Atomic Energy*, Geneva, **16** (1958), 535–549.

G. M. Wing, "On a Method for Obtaining Bounds on the Eigenvalues of Certain Integral Equations," *J. Math. Anal. Appl.* **11** (1965), 160–175.

G. M. Wing, "Mathematical Methods Suggested by Transport Theory," Edited by R. E. Bellman and G. Birkhoff, Vol. **1**, *Proc. Symposia on Applied Mathematics* (SIAM-AMS) American Math Society 1969, pp 159–177.

IX

NONLINEAR BOUNDARY-VALUE PROBLEMS

1   Introduction

The method of quasilinearization has been used very effectively for solving certain classes of nonlinear boundary-value problems for several years. The technique proceeds by replacing the nonlinear boundary-value problem by a sequence of linear boundary-value problems which converge to the solution of the nonlinear equation. The historical procedure of solving the sequence of linear boundary-value problems normally involves solving an associated homogeneous equation and an associated inhomogeneous equation and then using the principle of superposition to obtain the solution of each linear problem.

Several problems arise with this type of approach. First, the original nonlinear problem may be unstable with respect to an initial-value type of integration procedure. Many times this is reflected by the fact that the linearized problems are also unstable. One method which avoids the problem of instability is to solve the linear problems by means of an implicit finite difference scheme as described in section 9 of Chapter II.

Another problem which often arises is that one may be interested only in the value of the function or its derivative at the ends of the interval and how these change as the interval length is varied. This very common situation occurs in transport theory, structural mechanics, and electrochemical theory of the fuel cell. Neither the superposition nor the finite difference scheme is suited to this type of analysis, since each time the interval length is changed, a new problem is presented.

In this chapter we wish to show how the techniques of invariant imbedding and quasilinearization may be combined to solve certain nonlinear boundary-value problems.

198

2   Classical Invariant Imbedding

Consider the system of nonlinear ordinary differential equations,

$$\dot{u} = f(z, u, v) \qquad \text{(IX.2.1a)}$$

$$-\dot{v} = g(z, u, v), \qquad \text{(IX.2.1b)}$$

subject to the two-point boundary conditions,

$$u(0) = 0, \quad v(x) = c. \qquad \text{(IX.2.1c, d)}$$

(In this section, the dot will denote differentiation with respect to z.) In order to indicate the dependence of the functions u and v upon the x and c, as well as upon z, we shall write, when necessary for clarity,

$$u = u(z, c, x), \qquad \text{(IX.2.2)}$$

$$v = v(z, c, x). \qquad \text{(IX.2.3)}$$

If we differentiate in (IX.2.1) with respect to c, we obtain

$$\dot{u}_c = f_u u_c + f_v v_c, \quad u_c(0) = 0, \qquad \text{(IX.2.4a, b)}$$

$$-\dot{v}_c = g_u u_c + g_v v_c, \quad v_c(x) = 1. \qquad \text{(IX.2.4c, d)}$$

Similarly, differentiation with respect to x yields

$$\dot{u}_x = f_u u_x + f_v v_x, \quad u_x(0) = 0, \qquad \text{(IX.2.5a, b)}$$

$$-\dot{v}_x = g_u u_x + g_v v_x, \quad \dot{v}(x, c, x) + v_3(x, c, x) = 0. \quad \text{(IX.2.5c, d)}$$

In the boundary condition given in (IX.2.5d), $\dot{v}$ is the derivative of v with respect to its first argument and v_3 is the derivative with respect to its third argument.
Let us introduce the function $r(c, x)$ by the relation

$$r(c, x) = u(x, c, x). \qquad \text{(IX.2.6)}$$

Then (IX.2.5d) can be rewritten as

$$v_3(x, c, x) = -\dot{v}(x, c, x) = g(z, u, v)|_{z=x}$$

$$= g(x, r(c, x), c). \qquad \text{(IX.2.7)}$$

Comparing (IX.2.4) and (IX.2.5), we see that u_c, v_c satisfy the same linear differential equations as do u_x, v_x, except for the boundary conditions. By exercise 13 of Chapter III, we see that

$$u_x(z, c, x) = g(x, r(c, x), c) u_c(z, c, x), \qquad \text{(IX.2.8)}$$

$$v_x(z, c, x) = g(x, r(c, x), c) v_c(z, c, x) \quad 0 \leqslant t \leqslant x. \qquad \text{(IX.2.9)}$$

Equations (IX.2.8) and (IX.2.9) represent partial differential equations for u and v. The initial conditions are given by (IX.2.6) and

$$v(x, c, x) = c. \qquad \text{(IX.2.10)}$$

We must now obtain an equation for $r(c, x)$. If we differentiate in equation (IX.2.6) with respect to x, we obtain

$$r_x(c, x) = \dot{u}(x, c, x) + u_3(x, c, x). \qquad \text{(IX.2.11)}$$

From (IX.2.1 and (IX.2.8) evaluated at $z = x$, we have that

$$r_x(c, x) = f(x, r(c, x), c) + g(x, r(c, x), c) r_c(x, x). \qquad \text{(IX.2.12)}$$

This is the desired partial differential equation for $r(c, x)$. The initial condition is obtained from (IX.2.1c) and is

$$r(c, 0) = 0. \qquad \text{(IX.2.13)}$$

Let us summarize the results. For sufficiently small values of x, it is assumed that the function $r(c, x)$ is obtained as the solution of the Cauchy problem

$$r_x = f(x, r, c) + g(x, r, c) r_c, \qquad x > 0, \qquad \text{(IX.2.14a)}$$

$$r(c, 0) = 0. \qquad \text{(IX.2.14b)}$$

The function u is the solution of the problem

$$u_x = g(x, r, c) u_c, \qquad z < x \qquad \text{(IX.2.15a)}$$

$$u(z, c, x)|_{x=z} = r(c, x). \qquad \text{(IX.2.15b)}$$

Finally, the function v is obtained as the solution of the problem

$$v_x = g(x, r, c) v_c, \qquad z < x \tag{IX.2.16a}$$

$$v(z, c, x)|_{x=z} = c. \tag{IX.2.16b}$$

The essential feature of the method is that all of the equations are initial-valued, thus necessitating integration only in the direction of increasing x; that is, it is a one-sweep method.

3　Quasilinearization

Our attention will be centered around nonlinear ordinary differential equations of the form

$$L(y) = f(y, x), \tag{IX.3.1a}$$

with the boundary conditions

$$y(a) = y_a, \qquad y(b) = y_b, \tag{IX.3.1b}$$

where x is an element of the n-dimensional Euclidean space R^n. The operator L is a linear second-order differential operator which possesses certain positivity properties.

The $f(y, x)$ is continuous in y and x, is a strictly convex function of y for all x in a domain D, and has a bounded second partial derivative with respect to y for all y and x. Analogous results hold if the function is strictly concave. For simplicity, the x dependence of the function will not be specified unless needed for clarity.

A strictly convex function which is twice differentiable can be expanded by use of the mean value theorem as

$$f(y) = f(v) + (y - v) f'(v) + \tfrac{1}{2}(y - v)^2 f''(\theta_1), \tag{IX.3.2}$$

where $v \leqslant \theta_1 \leqslant y$. Since $f(y)$ is a strictly convex function of y, then

$$f''(y) > 0. \tag{IX.3.3}$$

Hence,

$$f(y) - [f(v) - (y - v) f'(v)] \geqslant 0, \tag{IX.3.4}$$

with equality holding for $y = v$. In terms of the maximum operation, (IX.3.4) can be written as

$$f(y) = \max_{v} [f(v) + (y - v) f'(v)].$$ \qquad (IX.3.5)

We now introduce the associated linear equation for the function $w = w(v, x)$, which is subject to the same boundary conditions as y. This leads us to the representation

$$L(w) = f(v, x) + (w - v) f_v(v, x),$$ \qquad (IX.3.6)

where f_v is the partial derivative of the function f with respect to v. We assume that the solution of (IX.3.6) and the original equation (IX.3.1) exists and is unique.

Theorem 1: If L is a linear operator possessing the positivity properties of (IX.3.8) and (IX.3.9) and $f(y)$ is a strictly convex function of y, for all finite y, then y, the solution of the nonlinear equation (IX.3.1), may be represented in the form

$$y = \max_{v} w(v, x),$$ \qquad (IX.3.7)

where $w = w(v, x)$ is the solution of the associated linear equation (IX.3.6).

Our positivity assumptions on L are the following: If z, which is zero on the boundary, satisfies the inequality

$$L(z) - f'(v) z \geqslant 0$$ \qquad (IX.3.8)

for all admissible functions, v, throughout the domain, D, then

$$z \geqslant 0.$$ \qquad (IX.3.9)

The establishment of the positivity property is sometimes very difficult. It is sufficient to say that many differential operators possess this property.

We now observe that (IX.3.5) can be written as

$$L(y) = f(y, x) + (y - v) f'(v, x) + q,$$ \qquad (IX.3.10)

where q is a nonnegative function. If we subtract (IX.3.6) from (IX.3.10), we obtain

$$L(y - w) = (y - w) f'(v) + q.$$ \qquad (IX.3.11)

Thus, we write

$$L(y - w) - (y - w) f'(v) \geqslant 0. \qquad \text{(IX.3.12)}$$

Since L satisfies the positivity properties of (IX.3.8) and (IX.3.9), the following inequality holds:

$$y \geqslant w. \qquad \text{(IX.3.13)}$$

If $v = y$, then

$$w = w(y, x) = y, \qquad \text{(IX.3.14)}$$

and it follows that

$$y = \max_{v} w(v, x). \qquad \text{(IX.3.15)}$$

Now we would like to construct a sequence of functions that converge monotonically to $y(x)$. We first choose an initial approximation $v_0(x)$ and use it to determine the function $w = y_0$ from the equation

$$L(y_0) = f(v_0) + (y_0 - v_0) f'(v_0), \qquad \text{(IX.3.16)}$$

subject to the original boundary conditions. Then the new function, y_0, is used to determine an improved admissible function, v_1, which maximizes the expression $f(v) + (y_0 - v) f_v(v)$ throughout the entire range of x. Since the function f is convex, this function is $v = v_1 = y_0$. We now use $v_1 = y_0$ and compute an improved value of y_1 from the equation

$$L(y_1) = f(y_0) + (y_1 - y_0) f'(y_0). \qquad \text{(IX.3.17)}$$

An improved admissible function, $y = y_2$, is then determined as the function which maximizes the expression $f(v) + (y_1 - v) f_v(v)$, which is $v = v_2 = y_1$. Continuing in this manner, we obtain the sequence of functions $\{y_n\}$ defined by the equations

$$L(y_0) = f(v_0) + (y_0 - v_0) f'(v_0),$$
$$L(y_{n+1}) = f(y_n) + (y_{n+1} - y_n) f'(y_n), \quad n = 1, 2, \dots, \qquad \text{(IX.3.18)}$$

which are subject to the original boundary conditions. The actual application of the method is quite simple and straightforward.

4   Quasilinearization and Invariant Imbedding

As we have seen in section 2, when the imbedding is applied to nonlinear problems, the resulting equations are nonlinear partial differential equations. Frequently, these equations appear to be at least as difficult to handle as the original problem, and the advantages that the method provides in treating linear problems are no longer present.

To circumvent this difficulty and the stability problems discussed in the previous section, we combine the techniques of invariant imbedding and quasilinearization. We use the latter device to reduce certain classes of nonlinear problems to sequences of linear problems. The invariant imbedding is then applied to each linear member of the sequence. Under appropriate conditions the corresponding sequence of solutions approaches the solution of the original problem.

We consider the two-point boundary-value problem

$$\frac{du}{dz} = f(z, u, v)$$

$$-\frac{dv}{dz} = g(z, u, v)$$

(IX.4.1a)

$$u(0) = \alpha, \qquad v(x) = \beta, \qquad 0 \leqslant z \leqslant x.$$

(IX.4.1b)

If (IX.4.1) is quasilinearized as in section 3, (IX.4.1) is replaced by

$$\frac{du_{n+1}}{dz} = a_n(z) u_{n+1} + b_n(z) v_{n+1} + e_n(z),$$

$$-\frac{dv_{n+1}}{dz} = c_n(z) u_{n+1} + d_n(z) v_{n+1} + f_n(z),$$

(IX.4.2a)

$$u_{n+1}(0) = \alpha, \qquad v_{n+1}(x) = \beta, \qquad 0 \leqslant z \leqslant x,$$

(IX.4.2b)

where

$$a_n(z) = f_u(z, u_n, v_n),$$

$$b_n(z) = f_v(z, u_n, v_n),$$

$$c_n(z) = g_u(z, u_n, v_n),$$

$$d_n(z) = g_v(z, u_n, v_n),$$

$$e_n(z) = f(z, u_n, v_n) - u_n f_u(z, u_n, v_n) - v_n f_v(z, u_n, v_n),$$

$$f_n(z) = g(z, u_n, v_n) - u_n g_u(z, u_n, v_n) - v_n g_v(z, u_n, v_n).$$

As we have already mentioned, for a significant class of problems the functions u_n and v_n converge to the solutions u and v of (IX.4.1), provided the u_0 and v_0 are appropriately chosen.

At this point we have several options available, since we may use any of the algorithms discussed in Chapter IV. For example, if we use the basic algorithm of section 2, of Chapter IV, the invariant imbedding functions satisfy

$$r'_{n+1}(z) = b_n(z) + [a_n(z) + d_n(z)] r_{n+1}(z) + c_n(z) r^2_{n+1}(z), \qquad \text{(IX.4.3a)}$$

$$s'_{n+1}(z) = [a_n(z) + c_n(z) r_{n+1}(z)] s_{n+1}(z) + f_n(z) r_{n+1}(z) + e_n(z), \text{(IX.4.3b)}$$

$$q'_{n+1}(z) = [d_n(z) + c_n(z) r_{n+1}(z)] q_{n+1}(z), \qquad \text{(IX.4.3c)}$$

$$t'_{n+1}(z) = [f_n(z) + c_n(z) t_{n+1}(z)] q_{n+1}(z), \qquad \text{(IX.4.3d)}$$

where we have used $r=r_1$, $s=r_2$, $q=q_1$, and $t=q_2$ in order to avoid obvious notational difficulties.

The solutions $u_{n+1}(z)$ and $v_{n+1}(z)$ are then found from

$$u_{n+1}(z) = r_{n+1}(z) v_{n+1}(z) + s_{n+1}(z), \qquad \text{(IX.4.4a)}$$

$$v_{n+1}(0) = q_{n+1}(z) v_{n+1}(z) + t_{n+1}(z). \qquad \text{(IX.4.4b)}$$

Thus, for each value of n, we must complete (IX.4.3, IX.4.4).

The choice of u_0 and v_0 is left open. Rather obvious is the possibility

$$u_0(z) \equiv \alpha,$$
$$v_0(z) \equiv \beta. \qquad \text{(IX.4.5)}$$

Partial knowledge of the solution of (IX.4.1) or some of its properties may be helpful in selecting u_0 and v_0. Indeed, such knowledge may be vital in obtaining convergence.

5 Numerical Examples

In this section we present three examples to illustrate the numerical technique described in the previous section. We begin with a brief discussion of the numerical procedures involved in implementing this method. The algorithm used was that described in excercise 1 of Chapter IV. An IBM 360 Model 44, using double-precision arithmetic, was used for the numerical integration with a step size varying from 0.001 to 0.01. The choice of stepsize depends upon the particular problem.

Past experience has shown that the imbedding method is most useful when applied to a family of problems of different "lengths" x. In our examples we, therefore, introduce the family

$$\frac{du}{dz} = f(u, v, z),$$

$$-\frac{dv}{dz} = g(u, v, z),$$

(IX.5.1)

$$u(0) = a, \quad v(x_i) = y, \quad i = 1, 2, ..., N,$$

$$0 < x_1 < x_2 < \cdots < x_N.$$

Denote the solution of the system (IX.5.1) on $[0, x_i]$ by $u^i(z)$, $v^i(z)$. For $i=1$ the sequences $\{u_n^1(z)\}$ and $\{v_n^1(z)\}$ were generated by using the initial iterates $u_0^1(z)$ and $v_0^1(z)$ as given by (IX.4.5). For $i=k$, $2 \leqslant k \leqslant N$, the sequences $\{u_n^k(z)\}$ and $\{v_n^k(z)\}$ were generated by using

$$u_0^k(z) = \begin{cases} u^{k-1}(z), & 0 \leqslant z \leqslant x_{k-1} \\ u^{k-1}(x_{k-1}), & x_{k-1} \leqslant z \leqslant x_k, \end{cases}$$

(IX.5.2a)

$$v_0^k(z) = \begin{cases} v^{k-1}(z), & 0 \leqslant z \leqslant x_{k-1} \\ y, & x_{k-1} \leqslant z \leqslant x_k. \end{cases}$$

(IX.5.2b)

For $x_k - x_{k-1}$ small, then $u_0^k(z)$ and $v_0^k(z)$, as defined in (IX.5.2), were quite close to the solutions $u^k(z)$ and $v^k(z)$, respectively, and the iterates converged quite rapidly.

We now proceed to discuss our examples.

Example 1

Perhaps the best classical nonlinear equation on which to try our numerical method is

$$\frac{d^2\psi}{dz^2} = e^\psi.$$

We pose this in the form

$$\frac{du}{dz} = v,$$

$$-\frac{dv}{dz} = -e^u,$$

(IX.5.3)

$$u(0) = 0, \quad v(x) = y.$$

The analytic solution is given implicitly by

$$\sqrt{2z} - \frac{2}{c} \log\left\{(c + \sqrt{1 + c^2})\left(\frac{e^{u(z)/2}}{c + e^{u(z)} + c^2}\right)\right\} = 0 \qquad \text{(IX.5.4a)}$$

and

$$\sqrt{2z} - \frac{2}{|c|}\left\{\cos^{-1}\left(\frac{|c|}{e^{u(z)/2}}\right) - \cos^{-1}|c|\right\} = 0. \qquad \text{(IX.5.4b)}$$

Here c is, of course, determined in each case by using the boundary condition $v(x)=y$ and the relationship

$$\tfrac{1}{2}v(x)^2 = e^{u(z)} + c^2,$$

which may be obtained by a simple manipulation using (IX.5.3).

The numerical results obtained for $y=0$ are given in Table IX.5-1. (For convenience we give only the values of u and v at the boundary points, $z=0$ and $z=x$.)

TABLE IX.5-1.   *Numerical Results for Missing Boundary Conditions as a Function of Interval Length*

x	$u(x)$	$v(0)$
0.1	−0.0050	−0.0997
0.2	−0.0197	−0.1974
0.3	−0.0434	−0.2915
0.4	−0.0751	−0.3805
0.5	−0.1137	−0.4636
0.6	−0.1578	−0.5404
0.7	−0.2063	−0.6106
0.8	−0.2581	−0.6745
0.9	−0.3123	−0.7324
1.0	−0.3681	−0.7848

These results were checked against the analytical solution given by (IX.5.4), and 4-digit agreement was obtained. The computation in Table IX.5-1 took about 3.4 minutes of computing time. This compares very favorably with other techniques available.

Example 2

As a second example, we examine

$$\frac{du}{dz} = v - \varepsilon uv,$$

$$-\frac{dv}{dz} = u - \varepsilon uv,$$

(IX.5.5)

$$u(0) = 0, \quad v(x) = y, \quad \varepsilon \geqslant 0.$$

Equation (IX.5.5) describes a simple transport process involving binary fission

TABLE IX.5-2.   *Numerical Results for Missing Boundary Conditions as a Function of Interval Length for $y=5$ and $\varepsilon=0.01$*

x	$u(x)$	$v(0)$
1.0	7.1197	8.6321
1.1	8.6395	9.9049
1.2	10.6016	11.6356
1.3	13.1988	14.0201
1.4	16.6703	17.3028
1.5	21.2172	21.6913

TABLE IX.5-3.   *Numerical Results for Missing Boundary Conditions as a Function of Interval Length for $y=100$ and $\varepsilon=0.01$*

x	$u(x)$	$v(0)$
1.0	76.5862	137.7468
1.1	79.3445	135.1366
1.2	81.7244	132.6055
1.3	83.7870	130.2344
1.4	85.5818	127.9925
1.5	87.1498	125.8811

and particle-particle interaction. If $\varepsilon=0$, the system becomes critical for $x=\pi/2$; if $\varepsilon>0$, there is no critical length, regardless of the input y.

For $y=5$ and $\varepsilon=0.01$, we obtained the results shown in Table IX.5-2, for $y=100$ and $\varepsilon=0.01$; data are given in Table IX.5-3

Example 3

Finally, to test the method on an ill-behaved problem, we chose

$$\frac{du}{dz} = v,$$

$$-\frac{dv}{dz} = -2u\,(1+u^2), \tag{IX.5.6}$$

$$u\,(0) = 0, \quad v\,(x) = \sec^2 x, \quad 0 \leqslant z < \pi/2.$$

The solution to (IX.5.6) is

$$u\,(z) = \tan z,$$

$$v\,(z) = \sec^2 z.$$

Clearly, for x near $\pi/2$, the problem presents a numerical challenge. Physically, we may think of $\pi/2$ as the "critical" length of a nonlinear transport problem; however, there has been no effort to develop a realistic transport analogy.

The data in Table IX.5-4 were generated by using an integration step size of 0.005. The scheme defined by (IX.5.2) was used to start each iteration $(k \geqslant 2)$. The number of iterations which yielded a uniform pointwise error of less than 5×10^{-6} is recorded as a matter of interest.

TABLE IX.5-4. *Comparison of Invariant Imbedding Solution and Exact Solution of $u(x)$ as a Function of Interval Length*

x	$u(x)$	$\tan x$	Number of Iterations
1.0	1.5574	1.5574	5
1.3	3.6022	3.6021	5
1.5	14.1073	14.1014	8
1.51	16.4654	16.4281	5
1.52	19.6856	19.6695	5
1.53	24.5293	24.4984	5
1.54	32.5326	32.4611	5
1.55	48.2975	48.0785	6

6 Exercises

1. Prove that the algorithm presented in section 2 does indeed solve the original problem.

2. Consider the equation

$$x + bx^n = a, \quad a, b > 0, \quad n > 1. \tag{1}$$

Using the maximum operation show that

$$x^n = \max_{y \geqslant 0} [nxy^{n-1} - (n-1) y^n], \quad x \geqslant 0, \quad n > 1 \tag{2}$$

The maximum is attained for $y = x$.

3. Substitute (2) into (1) and show that

$$x \leqslant \frac{a + b(n-1) y^n}{1 + bny^{n-1}}.$$

Since equality is attained at $y = x$, we may write the explicit analytic representation for the solution as

$$x = \min \frac{a + b(n-1) y^n}{1 + bny^{n-1}}.$$

4. If the boundary conditions are nonlinear, they must also be linearized. For example, consider the boundary conditions

$$f(u(0), u'(0)) = 0,$$
$$g(u(x), u'(x)) = 0.$$

Linearize these equations and show that the new boundary conditions are of the form

$$\alpha_n u_{n+1}(0) + \beta_n u'_{n+1}(0) = \eta_n$$
$$\gamma_n u_{n+1}(x) + \delta_n u'_{n+1}(x) = \varepsilon_n.$$

7   Bibliographical Discussion

Section 2

There are a number of references dealing with application of the classical invariant imbedding to nonlinear boundary value problems. A few of these are

R. E. Bellman, R. E. Kalaba and G. M. Wing, "Invariant Imbedding and the Reduction of Two-Point Boundary-Value Problems to Initial Value Problems," *Proc. Nat'l. Acad. Sci. USA* **46** (1960), 1646–1649.

R. E. Bellman and R. E. Kalaba, "On the Fundamental Equations of Invariant Imbedding, I," *Proc. Nat'l. Acad. Sci. USA* **47** (1961), 336–338.

M. A. Golberg, "Some Invariance Principles for Two-Point Boundary-Value Problems," *J. Math. Anal. Appl.* **36** (1971), 141–148.

M. A. Golberg, "A Generalized Invariant Imbedding Equation, I," *J. Math. Anal. Appl.* **33** (1971), 518–528.

M. A. Golberg, "A Generalized Invariant Imbedding Equation, II," *J. Math. Anal. Appl.* **34** (1971), 509–601.

M. A. Golberg, "A Generalized Invariant Imbedding Equation, III," *J. Math. Anal. Appl.* **35** (1971), 209–214.

H. H. Kagiwada and R. E. Kalaba, "Derivation and Validation of an Initial Value Method for Certain Nonlinear Two Point Boundary Value Problems," *J. Opt. Th. Appl.* **2** (1968), 378–385.

Section 3

The best discussions of quasilinearization are found in

R. E. Kalaba, "On Nonlinear Differential Equations, the Maximum Operation, and Monotone Convergence," *J. Math. Mech.* **8** (1959), 519–574.

R. E. Bellman and R. E. Kalaba, *Quasilinearization and Nonlinear Boundary Value Problems*, American Elsevier Publishing Co., New York, 1965.

Sections 4 and 5

The material of these sections first appeared in

R. C. Allen, Jr., M. R. Scott and G. M. Wing, "On the Numerical Solution of Certain Nonlinear Boundary Value Problems," *J. of Comp. Phys.* **4** (1969), 250–257.

AUTHOR INDEX

SUBJECT INDEX